NEW-AGE TECHNOLOGY AND INDUSTRIAL REVOLUTION 4.0

Also by The Same Author

Bharat Ratna Dr Babasaheb Ambedkar
An Intellectual Colossus, Great National Leader and Universal Champion of Human Rights (Photo-Biography in English and Marathi)

Ambedkar *An Economist Extraordinaire*

Ambedkar *Awakening India's Social Conscience* (An Intellectual Biography) (English and Marathi)

Ambedkar Writes
Completed Writings of Dr Ambedkar (Edited) (English, Marathi and Hindi)
Vol. I: Political Writings, Vol. II: Scholarly Writings

Ambedkar Speaks: *Seminal Speeches*
Edited: Vol. I, II and III (English 300, Marathi 500)

Trilogy on Ravindranath Tagore (Marathi)
Ravindranath Tagore : Yuga Nirmata Vishvamanav
Ravindranath Tagore : Samagra Sahitya Darshan
Bhaya Shoonya Chitta Jeth : 151 Pratinidhik Kavita

Untouchables
My Family's Triumphant Journey Out of the Caste System in Modern India (International Bestseller – in 15 languages)

Monetary Policy, Financial Stability and Central Banking in India

Re-emerging India – A Global Perspective

Challenges to Indian Banking: Competition, Globalization and Financial Markets (Edited)

Monetary Economics for India

Aamcha Baap Aan Amhi (Marathi, Family Autobiography)
The largest selling book in the history of Marathi literature

NEW-AGE TECHNOLOGY AND INDUSTRIAL REVOLUTION 4.0

Global Public Policy Issues in Economy, Democracy, National Security and World Peace

Dr Narendra Jadhav

KONARK

Konark Publishers Pvt Ltd
New Delhi • Seattle

Konark Publishers Pvt Ltd
206, First Floor,
Peacock Lane, Shahpur Jat,
New Delhi - 110 049
Tel : +91-11-4105 5065
Mob : +91-93552 93900, +91-93552 94900
e-mail : india@konarkpublishers.com
website : www.konarkpublishers.com

Konark Publishers International
8615, 13th Ave SW,
Seattle, WA 98106
Phone : (415) 409-9988
e-mail : us@konarkpublishers.com

First Impression 2019

Cataloging in Publication Data--DK
Courtesy: D.K. Agencies (P) Ltd. <docinfo@dkagencies.com>

Jadhav, Narendra, 1953- **author.**
New-Age technology and industrial revolution 4.0 : global public policy issues in economy, democracy, national security and wold peace / Dr Narendra Jadhav.
pages cm
Includes index.
ISBN 9789322008994

1. Technological innovations. 2. Technological innovations--India.
3. Industrial revolution. 4. Industrial revolution--India. 5. Political planning.
6. Political planning--India. I. Title.

LCC HC440.T4J33 2019 | DDC 338.0640954 23

Cover jacket and book designed by idesign
Printed and bound in India by Lustra Print Process, New Delhi

To

The Three Great Doctors

Dr Suresh Advani, Dr Raman Deshpande and Dr Anil Bhoraskar
who pulled me out from the jaws of Cancer
and gave me a new lease of life
in
Gratitude
forever.

CONTENTS

PART • FOUR
A CASE-STUDY

AUTHOR'S NOTE

The rationale for this book comes straight from the latest G-20 declarations. The Group of Twenty (G-20) is perhaps the most powerful group of systemically important Advanced countries as well as developing economies—from the US to China and India.

The G-20 leaders met in Osaka, Japan on 28-29 June 2019 and issued a Declaration. This historic Declaration *inter alia* states:

i) We, the Leaders of the G-20, met in Osaka, Japan on 28-29 June 2019 to make united efforts to address major global economic challenges. We will work together to foster global economic growth, while harnessing the power of technological innovation, in particular digitalization, and its application for the benefit of all (Para 1).

ii) We will work toward achieving an inclusive, sustainable, safe, trustworthy and innovative society through digitalization and promoting the application of emerging technologies (Para 10).

iii) As digitalization is transforming every aspect of our economies and societies, we recognize the critical role played by effective use of data, as an enabler of economic growth, development and social well-being. We aim to promote international policy discussions to harness the full potential of data (Para 10).

iv) Cross-border flow of data, information, ideas and knowledge generates higher productivity, greater innovation, and improved sustainable development, while raising challenges related to privacy, data protection, intellectual property rights, and security (Para 11).

This book is all about addressing precisely the same set of issues which are underscored by the G-20, as concerns of great contemporary relevance.

This book is about the emerging New-Age Technology underlying the rapidly unfolding Industrial Revolution 4.0, its impact on economy and society, and implications for public policy in a global setting.

Part One of the book presents a primer on the New-Age Technology and its constituents, including Artificial Intelligence and Machine Learning, Big Data Analytics, Robotics, Blockchain, Automated Technology, Additive Manufacturing, Internet of Things and Augmented Reality.

Part Two of the book discusses potential impact of the New-Age Technology on economic growth, employment, socio-economic inequalities and human development besides the evolving Banking-Finance 4.0.

Part Three of the book dilates on the associated global public policy issues, including Right to Privacy, social media and democratic governance, cybersecurity, surveillance and threats to national security and world peace arising from weaponization of the technology. This part is roundedup by discussion of issues relating to regulation and strategic development of the New-Age Technology.

While the entire book is in a global setting, the final part of the book includes a brief chapter on India as a case study.

This book, probably the first of its kind, bringing it all together—technology, economy, society and national security, would benefit individuals, businesses, corporates, civil society as well as policymakers around the world.

Narendra Jadhav

ACKNOWLEDGEMENTS

Research Support

The Dialogue team led by Kazim Rizvi – Founding Director

Preliminary drafts provided by Harsh Bajpai, Pranav Bhaskar Tiwari, Ayush Tripathi, Paranjay Tripathi, Mrittika Guha Sarkar and Kazim Rizvi

Logistics and plagiarism check by Kazim Rizvi
Also Vikas Mahajan, Pune

Editorial Assistance

Anna Chandy, Jiza Joy and Lekshmi Parameswaran

Technical Assistance

Sudha Rajagopalan, Prabhakar Gaikwad, Torsha Dasgupta, Snehal P. and Simran Sharma

Other Assistance

Deepak Negi, Rajesh Kumar, Samir Arora, Abhimanue Tiwari, Lalu Yadav and Pradeep Kumar

Family Support

Vasundhara Jadhav (Wife), Tanmoy (Son), Dr. Apoorva (Daughter)

Kejal (Daughter-in-law), Dr Ekim Muyan (Son-in-law)

Agastya, Eymir, Kimaya and Keyan (Grandchildren)

1

Introduction

New-Age Technology: Future Fast Forward

"Humanity today is like a waking dreamer, caught between the fantasies of sleep and the chaos of the real world..... We have created a Star Wars Civilization, with Stone Age emotions, Medieval institutions, and God-like technology.... we are terribly confused...."

Edward O. Wilson[1]
Biologist and Naturalist

When Pulitzer Prize winning American author Edward O. Wilson made this statement, he was not alone. Several great minds such as Stephen Hawking, Steve Wozniak (co-founder of Apple) and Elon Musk (founder of Tesla and SpaceX) and more than a thousand high-profile researchers wrote an open letter, way back in 2015, warning the world that Artificial Intelligence (AI) can potentially be even more dangerous than nuclear weapons.[2] Elon Musk, a leading technology entrepreneur himself, went to the extent of declaring Artificial Intelligence (AI) as the most serious threat to the survival of the human race.[3]

In a way, this book is an attempt to understand, critically examine and elaborate on at least some of the many dimensions of these thought-provoking assertions, especially those relating to the so-called Industrial Revolution 4.0.

The book discusses the emerging New-Age Technology underlying the Industrial Revolution 4.0, brings out opportunities and threats arising there from and makes an attempt to identify and analyze relevant public policy issues in the global context. The book maintains a global and cross-country perspective throughout.

In this introductory chapter, Section I presents an exceedingly brief overview of the evolution of technology from time immemorial. Successive Industrial Revolutions are covered right upto the ongoing Industrial Revolution 4.0. Against that background, Section II discusses the potential of the Industrial Revolution 4.0 in terms of opportunities for economic growth and human development. That is contrasted against the challenges posed by the emerging New-Age Technology raising complex issues and concerns relating to economy, society, democracy, national security and global peace. The final section, i.e. Section III provides the details of the proposed structure of this book.

I. Historical Evolution of Technology

From the beginning of history, technology has been the prime mover in shaping the evolution of human society. In fact, most theories of social and cultural evolution of human beings acknowledge that technological progress has been the primary factor driving the development of human civilization.

A pioneering American Anthropologist Professor Lewis H. Morgan (1818-1881), argued way back in the nineteenth century that social evolution of human beings from savagery through barbarism to civilization can be divided simply on the basis of technological milestones beginning with the creation of fire and its manipulation.[4]

A later day American Anthropologist Professor Leslie White (1900-1975), who championed the cause of 'neoevolutionism' differentiated human development between five stages, on the basis of 'harnessing and controlling energy'. In stage one, people used the energy of their own muscles; in the second stage, they used the energy of domesticated animals; in the third, the energy of plants (i.e. agriculture) was used; next, people learnt to use the energy of natural resources, such as coal, oil and gas; and finally, they learnt to harness the nuclear energy. The underlying premise of this classification was that "culture evolves as the amount of energy harnessed per capita per year is increased or as the efficiency of the instrumental means of putting the energy to work is increased".[5]

While maintaining the one-to-one correspondence between evolution of technology and human society, an American Sociologist Gerhard Lenski (1925-2015) focused on information. According to him, the more information and knowledge a given society has (in terms of shaping and manipulating its natural environment), the more advanced it is.[6]

Lenski discussed four stages of human development: first, when information is passed on by genes; secondly, when human beings learn and pass information through experience; thirdly, when human beings started using signs and developed logic; and, fourthly, when they created symbols, developed language and writing skills.

According to Lenski's "ecological-evolutionary theory", human societies could be divided chronologically into five types, on the basis of their level of technology, communication and economy:

- Hunting and Gathering Societies
  - existed since the beginning of human life
  - used primitive weapons
- Horticulture and Pastoral Societies
  - existed since 12000 years ago
  - used hand tools
- Agrarian Societies
  - existed since 5000 years ago
  - used animal-drawn plough

- Industrial Societies
  - in existence since about 1750
  - characterized by mechanized production using advanced sources of energy
- Post-Industrial Societies[7]
  - in existence roughly from 1980s
  - characterized by importance of services and information rather than industry and merchandise.

The Industry-era (1750 to date) is divided into the four Industrial Revolutions.

First Industrial Revolution (1750-1870) was spurred on by the use of steam powered engines and was characterized by the development of iron and textile industries. This Revolution began the process of turning the mostly agrarian and rural habitats into urban industrial societies.

The Second Industrial Revolution (1870-1914) was engineered when steam was replaced by water and electricity enabling mass production. This Revolution was characterized by expansion into new industries like steel, oil and electricity, and new products such as light bulbs, telephone and phonograph.

The Third Industrial Revolution (1980 onwards) was engendered by information technology. This Revolution was characterized by inventions such as the personal computer (PC) and the Internet besides information and communications technology (ICT). Since this Revolution entails a progression from analog electronic and mechanical devices to digital technology, it is also called the Digital Revolution.

The Industrial Revolution 4.0 is currently underway. According to Klaus Schwab (Founder and Executive Chairman of World Economic Forum), the distinguishing feature of the Industrial Revolution 4.0 is that its underlying basis lies not in advances in technology, but in breakthroughs in communication and connectivity.[8]

According to Dr Raghunath Mashelkar, Industrial Revolution 4.0 involves "the confluence of several exponential technologies like Robotics, Artificial Intelligence (AI), Machine Learning (ML), Big Data, etc."[9] As clarified, exponential technologies are those whose performance improves exponentially without any additional cost. Classic examples are Moore's law which states that computing power doubles every two years and Kryder's Law which observes that the disc drive density doubles every 13 months, i.e. storage becomes cheaper.

Since it is the combinatorial effect of diverse technologies and the resultant synergy that characterizes the Industrial Revolution 4.0, the emerging technology may be called post-Digital Revolution Fusion Technology.

The post-Digital Revolution Fusion Technology (mostly referred to as "the New-Age Technology" or simply 'the emerging revolutionary technology' in this book) is currently unfolding and is going to change our lives as never before.

II. New-Age Technology: Opportunities and Threats

Not too long ago, the Digital world that we live in today, was considered a matter of science fiction. As Bhaskar Pramanik, former Chairman of Microsoft (India) puts it: "We stand on the cusp of a new era of technology".[10] At the core of this New-Age Technology is cloud computing, which is a vast network of distributed data centres, providing for processing of data at unprecedented scale and speed.

"Mastering the Fourth Industrial Revolution" was the theme of the Word Economic Forum (WEF) Annual Meetings 2016 at Davos, Switzerland. While announcing the opening of its new centre for the Fourth Industrial Revolution in San Francisco, USA, the WEF observed:

> The possibilities of billions of people connected by mobile device, with unprecedented processing power, storage capacity, and access to knowledge are unlimited. And these possibilities will be multiplied by emerging technology breakthrough in fields such as Artificial Intelligence, Robotics, the Internet of Things, Autonomous vehicles, 3D printing, Nanotechnology, Biotechnology, Material Science, Energy storage and Quantum Computing.[11]

What was deemed to be science fiction is fast becoming a reality.

While opportunities and possibilities are immense, the rapidly unfolding New-Age Technology has also been raising complex questions and far-reaching societal concerns.[12]

The emerging revolutionary technology as it is understood today, seems to have the following distinct potential advantages:

1. Labor productivity would increase sharply, leading to significant additional wealth generation while accelerating global economic growth.
2. Poverty could be drastically reduced, even eradicated if technological advances are combined with appropriate Public Policy measures.
3. In several sectors of the economy, unprecedented technological breakthroughs would be possible. These would include, *inter alia* agriculture, industry, services, financial sector, education and healthcare.
4. The emerging technology could be deployed to augment ongoing global efforts aimed at containing the fallout from global warming and climate change.
5. Digital Assistants available today such as Alexa and Siri could become far more capable in future. They could be trained to anticipate individual needs, help manage schedules and reply to and route communications besides, driving cars.

6. If mother Earth becomes uninhabitable either due to human-made problems (such as wide scale nuclear war or climate change) or natural disasters (like an asteroid strike), it is conceivable that using the New-Age Technology, new habitats could be created outside the Earth – either on our moon or some other planet and its moon(s).

At the other extreme, the same technological revolution could drive the human beings to a rather dark future. Several downsides of the emerging technology must be taken on board. These include the following:

1. While economic growth will get a big boost, there may not be a commensurate increase in jobs. In fact, some reports have indicated a massive displacement of jobs on account of automation, especially in countries like India and China.
2. While the global economy would be on a high growth trajectory, economic disparities and social inequities may rise, thereby weakening or even threatening the social fabric and undermining the functioning of democracy.
3. Individual or business security could be jeopardized on account of Ransomware attacks which are already rising in frequency and intensity. Typically, criminals hide links to their Ransomware in seemingly innocuous e-mails or web pages, which when activated prohibit users from opening their own files or operating applications or system. This seizure continues until a ransom is paid, generally in some anonymous Cryptocurrency, like Bitcoin. (A latest version of Ransomware is called "WannaCry" and that says it all.)
4. Yet another area of serious concern is the possibility of cyberwarfare. Attacking critical infrastructure (i.e. utilities such as water and power, finance, health and transportation) would be possible using laser beams from space. Anti-satellite missiles could be used to destroy enemy's surveillance and communications network.

In either case, the targeted country could be seriously paralyzed.

5. Finally, with the "man-machine" equation undergoing a complete metamorphosis, the possibility of machines taking over and making human beings as second-class citizens cannot be ruled out.

The great challenge going forward is reaching a global consensus on ethical principles and values to govern or at least regulate the development and use of the emerging revolutionary technology.

The future of the humankind depends on it, literally.

III. Structure of the Book

The book comprises four parts:

- Part One presents the "what and how" of the New-Age Technology underlying the Industrial Revolution 4.0. This part has three chapters:
  - One on Artificial Intelligence (AI) and Machine Learning (Chapter 2), another one on Big Data and Data Analytics (Chapter 3) and third one on remaining constituents of the emerging technology [i.e. Automated Technology, Internet of Things, Robotics, Additive Manufacturing, 3D Printing and Augmented Reality and Blockchain Technology] (Chapter 4).
- Part Two is focused on potential implications of the emerging technology. This part also consists of three chapters:
  - One on global economic growth and human development (Chapter 5), covering several sectors, including agriculture, manufacturing, services, infrastructure, education

and healthcare, another one (i.e. Chapter 6) on socio-economic inequalities and the third one (i.e. Chapter 7) on financial system of the future (including e-Commerce and cryptocurrencies).

- Part Three is devoted to global Public Policy issues in the context of serious concerns and anxieties regarding the New-Age Technology. This part has six chapters, covering Right to Privacy (Chapter 8), Social Media and Democratic Governance (Chapter 9), Cybersecurity (Chapter 10), Surveillance (Chapter 11), Weaponization of Technology, National Security and Global Peace (Chapter 12), and issues relating to Regulation and Strategic Development of the emerging revolutionary technology (Chapter 13).
- Part Four has only one chapter (i.e. Chapter 14) which presents a case-study of India. Development, application and India's policy stance in respect of New-Age Technology have been reviewed briefly.

PART • ONE

NEW-AGE TECHNOLOGY: AN OVERVIEW

2

Artificial Intelligence and Machine Learning

"The challenge and potential of exploring this co-evolutionary future (between human intelligence and artificial intelligence) is the biggest story of the next century and one, in which, a closeness in development velocity is a necessity."

Bryan Johnson
American Venture Capitalist

"It is not Artificial Intelligence that I am worried about; it is the human stupidity (that worries me)."

Neil Jacobstein[1]
Professor of AI

Introduction

Instinctive behavior is what ensures the survival of all living beings. Such behavior in humans is somewhat peculiar, not being limited only to ensuring survival, but also aimed at securing comfort. The phrase that 'man cannot live by bread alone', although a cliché, continues to remain valid. Whenever new technologies arrive they often make human lives more efficient and comfortable, going a step further ahead from the instinctive purpose of survival.[2] This is why Artificial Intelligence (AI), a great dream of the humankind through the late twentieth and early twenty-first centuries is now closer to reality than ever before. Narrow AI systems used only for specific purposes are already in place and here onwards, AI systems will only get more sophisticated.

Going forward, there are different possibilities on how widespread adoption of AI will impact the economies, societies and the nature of governance. On one hand, there is the worst case scenario where AI is seen as a fundamental threat to the very existence of humanity.[3] On the other hand, there is a scenario where AI and humans achieve symbiosis and collectively work towards a better future. Regardless of the path chosen by human beings, there is consensus that the role of AI is all set to grow in the future.[4] Policymakers all over the world must, therefore, keep pace with the advances in AI and strive to regulate it while remaining proactive rather than reactive, as much as is possible.[5]

This chapter, dealing with a phenomenon that is already bringing dramatic changes in human lives, has been organized in six sections as follows: Section I briefly recapitulates the history of AI. Section II presents some prominent definitions of AI that have been proposed so as to underscore why this phenomenon is so hard to define precisely. Section III draws similarities between the human intelligence and machine intelligence. Section IV delves into the potential contribution of harnessing the power of AI. In Section V, some identified limitations of AI are discussed. Section VI elaborates on the so-called AI control problem, which has been worrying many experts and thinkers as an "existential threat".

I. Brief History

"Artificial Intelligence" or "AI", unlike in the 1980s and 1990s, not only rings a bell for scientists and engineers, but also for common people. People may not be fully aware of the manifold advancements in data storage capabilities or computing and how it is relevant to AI. But most people already know the ways in which they have become increasingly dependent on technology over the past 10 years or so.

The story of Artificial Intelligence begins in the era of the Second World War. The Germans used codes generated by the so called "Enigma Machine" to communicate. The Allies found these codes almost impossible to break, as the permutations and combinations that Enigma created were numerous. Then came along Alan Turing. He invented a machine called the "Bombe Machine" which could decipher codes made by Enigma. It was Turing who laid the foundation for what we call today "Machine Learning".[6]

In 1958, John McCarthy, said to be the father of AI, developed a language of programming known as Lisp, which was suitable for Machine Learning. (Before that, in 1956, the term "Artificial Intelligence" was coined by McCarthy at a conference organized by him.) In 1966, Shakey the Robot was invented, and it was recognized as the "first electronic person", because it was mobile and could reason about its own actions. In 1997, Deep Blue became the first super computer to defeat the then World Chess champion, Garry Kasparov. More recently, in 2017, AlphaGo system defeated the world number one Go player KeJie. This was indeed a landmark achievement, since Go is a game where moves made by players are based on intuition.[7]

While discussing the history of AI it is pertinent to mention the so-called "AI winters". Developments in the field of AI have been sporadic throughout its history. There were periods of dramatic development followed by several uneventful periods. The years between 1974-1980 and 1987-1994 are known as "first winter" and "second winter", respectively, as these were the periods that lacked breakthroughs in the field of AI.[8]

In the ongoing era, computation and data storage capacities are better than ever before and this has brought AI closer to reality, if not into reality itself. However, given the historical pattern, nobody can tell if and when the next winter will come. Therefore, it is possible that people who suggest that we are close to developing general purpose AI may be far off the mark. An American AI Researcher Eliezer Yudkowsky who has been championing the cause of "friendly AI", once observed: "By far the greatest danger of Artificial Intelligence is that people conclude too early that they understand it."[9]

II. Meaning of Artificial Intelligence

Defining AI itself may be more difficult than inventing it. This is so because of its dynamic nature. To understand the phenomenon better, look at some of the popular definitions that have been proposed.

According to Nils J. Nilsson:

> Artificial intelligence is that activity devoted to making machines intelligent, and intelligence is that quality that enables an entity to function appropriately and with foresight in its environment.[10]

Now the question is whether every machine that has foresight and functions appropriately in its environment should be characterized as AI? Take the example of a typewriter; upon pressing a key on the typewriter, the relevant alphabet is imprinted. In this case the machine can be said to have a foresight as to which alphabet it must imprint once the key is pressed but does this qualify it as AI? Similar questions arise in the case of calculators, telephones, etc.

Marvin Minsky and John McCarthy have similar definitions of AI. According to Minsky, AI is "the science of making machines do things that would require intelligence if done by men" and according to McCarthy, AI is "making a machine behave in ways that would be called intelligent if a human was so behaving". Going back to the typewriter example, if a human were to perform the task that a typewriter performs, i.e. writing, the same would require some level of intelligence. Does that mean that a typewriter is artificially intelligent? A calculator is an even better example. It can calculate various arithmetic problems faster than most humans. If a human were to do the same task, he or she would have to be tremendously intelligent. Does that mean that calculators are artificially intelligent?

It follows that Artificial Intelligence is a dynamic phenomenon that varies with time, which is why defining it is an exceedingly difficult task.[11] Machines that may be considered artificially intelligent today may lose this status 20 to 30 years from now. The devices that may seem artificially intelligent today will be considered too simple to be called so in the future. This is referred to as the AI effect or the "Odd Paradox".[12]

III. AI *vis-à-vis* Human Intelligence

The way humans acquire knowledge is through logical discourse wherein claims are presented about certain phenomena and these have to go through rigorous tests for wider acceptability. For example, if a scientist comes up with a certain theory (claim) about the nature of subatomic particles, he or she will have to convince the relevant scientific community through discourse and tests that his/her theory is valid.

Typically, claims are made by using logic. There can be two types of logical claims: those based on *deductive* logic and those made using *inductive* logic.

Deductive reasoning implies that given the existence of certain premises, a conclusion can be reached definitively, i.e. if the said premises exist, no other outcome can possibly be inferred.[13] If the premises exist, so does the conclusion. The deductive logic essentially differentiates between valid and invalid arguments. If the premises lead to the conclusion, the argument is valid. However, if the premises fail to logically derive the conclusion, the argument becomes invalid.

A good way to explain deductive reasoning is through syllogisms.[14] Consider the following argument:

If all apples are oranges; (Premise 1)

And all oranges are mangoes; (Premise 2)

Then all apples are mangoes (Conclusion)

The above mentioned is a valid deductive argument since the premises definitively lead to the conclusion.

If all lions are cats; (Premise 1)

And all cats are dogs; (Premise 2)

Then all dogs are lions (Conclusion)

The above mentioned is an invalid argument because the premises do not definitively lead to the conclusion.

The utility of deductive logic lies in segregating valid arguments from invalid ones and is essential to most disciplines. For example, it plays a major role in proving mathematical theorems.[15]

On the other hand, inductive logic or reasoning is probabilistic, *not* conclusive. Inductive arguments are probabilistic conclusions drawn from observation of patterns, possibly with the help of data analysis. For example, a person throws a tennis ball at a wall with great force; the ball bounces back with high velocity. He throws the ball again, this time with considerably lesser force. He notices the decrease in velocity with which the ball comes back. The person does this multiple times, notices a pattern and comes to the conclusion that "every action has an equal and opposite reaction",[16] which is, of course, Newton's Third Law of Motion.

It needs to be emphasized that unlike deductive arguments, inductive reasoning is not conclusive and the inferences drawn are *probabilistic* in nature. Continuing with the above-mentioned example, the conclusion that "every action has an equal and opposite reaction" is probabilistic because the pattern observed is limited only to the extent of the amount of data collected. If the same exercise was conducted by the person in outer space or in a different dimension where nature acts differently, his conclusions may well have been different.

It may be noted that there is no cause and effect relationship drawn between throwing of the ball and its bouncing back. There is merely a pattern observed and an inference drawn. Since inductive arguments are probabilistic in nature, new data can always change the conclusion in the same way that scientific conclusions are proved and disproved, from time to time, as more data are analyzed.[17]

Evidently, inductive reasoning holds an important place in scientific acquisition of knowledge. In fact, common people also use this inductive reasoning method in their day-to-day life. For example, consider a person learning to drive a car. The first time he or she presses a foot on the accelerator, and inadvertently presses it hard, the car moves faster than expected. Thereafter, the person presses the accelerator softly and finds that the car is moving too slowly. The person then presses it firmly and the car picks up a pace. Practice the same multiple times, and the person gets the hang of how to use the accelerator. This is an example of the inductive method of learning.

Now, the interesting phenomenon here is that the way machines learn is very similar to the way humans learn. In other words, machine learning and human learning seem almost analogical. To understand how this is possible, it would be instructive to see how machines learn.

A. Machine Learning

Machine Learning is an essential part of building AI. While using applications such as Facebook, YouTube, Amazon, Netflix, etc., interaction with machine-learning AI becomes inevitable. Machine Learning essentially involves statistical methods to analyze data. This means creating algorithms that enable machines to analyze data and make a determination or a prediction from the same.[18] This is known as *Unsupervized Machine Learning.* To an extent, this is analogous to inductive reasoning, in as much as its results are probabilistic.

Under *Supervized Machine Learning* on the other hand, a machine is consciously taught the conclusion it has to reach, and the same is not learnt by the machine on its own. For example, if the aim is to teach the machine to differentiate between apples and coconuts, it may be taught to differentiate by using colour as a parameter. So, when the machine detects brown colour, it concludes that the object is a coconut.[19] When it detects red colour, it concludes that the object is an apple. This is strikingly similar to deductive reasoning as far as differentiation between apples and coconuts is concerned.

B. Deep Learning

Deep Learning is a subset of Machine Learning. As a concept, it is almost congruent to the inductive method of learning in humans. Deep Learning uses artificial neural networks by which it teaches itself to achieve a particular task by experience. Deep Learning requires huge amounts of data. In recent years data storage capacity has increased manifold. Therefore, the feasibility of successful development of Artificial Intelligence by Deep Learning has also increased.[20]

In Deep Learning, a machine learns from the mistakes it commits and improves itself. This is very similar to the human learning process. For example, when a person starts playing a new video game, initially he or she does not play very well and loses the game easily. But as he or she slowly progresses and plays the game multiple times, the person learns to avoid mistakes made previously.

AlphaGo, the Deep Learning AI works exactly in the same manner. What is more, AlphaGo may even have outsmarted human intuition through Deep Learning (Go is an ancient Chinese game which in several instances requires intuitive moves. This is what makes Go different from games like Chess where a machine can learn to play by algorithm *per se*, rather than experience). So if one were to ask a good Go player as to why he or she made a certain move, the person won't reveal any strategy behind that move except the fact that the move "felt" right. The game may be the epitome of inductive method of learning. As mentioned earlier, AlphaGo defeated the Go World Champion Lee Seedol in 2016.

AlphaGo falls in the category of General Purpose AI, which means that it does not have a specific predefined purpose and can play any game that comes its way, unlike AI such as Siri or Okay Google which have a narrower purpose. The way AlphaGo works is this: that it is given a game to play which it plays multiple times, learns from its mistakes, improves itself and consequently plays better in subsequent attempts.

This indeed is starkly similar to the human way of learning and exceedingly fascinating because in some ways it helps us peep into the future of AI and how similar it may be to human intelligence.[22]

Of course learning from one's mistakes is optional for humans. Don't we know enough people who go on repeating their past mistakes?

IV. Potential Contributions of AI

Artificial Intelligence has tremendous applications in a very wide range of human endeavours. It is illustrated in the global context briefly, with three areas: Manufacturing, Healthcare and Space Exploration.

A. Manufacturing

There is a global trend of using new machines to replace humans in doing tasks that are too cumbersome for human bodies, thus, making manufacturing more efficient. Apart from handling difficult tasks, AI can even ensure precision in manufacturing by identifying minute defects in products that may not have been visible to the human eye. With AI at the horizon, these machines are bound to become intelligent apart from being capable of doing heavy tasks. This is expected to raise profits considerably in the manufacturing sector worldwide.[23] Further, AI will simultaneously create new jobs. It can even be harnessed to keep supply chains more efficient by predicting demand, taking into consideration factors such as macroeconomic cycles, political developments and weather patterns.[24]

B. Healthcare

An exponentially expanding ability to analyze huge amounts of data means quicker acquisition of knowledge. This becomes critical for research in disciplines such as medical science that are important for human existence. AI can analyze complex and large amounts of data with greater efficiency to reach a conclusion. Further, AI can be used to search literature from various studies that were undertaken previously to aid discovery of various drugs. According to research conducted by California Biomedical Research Association, it takes nearly 12 years to develop a drug. Only five out of 5,000 drugs that begin pre-clinical testing actually qualify for human testing, and only 20 per cent of those qualified are ever approved for human usage. This whole process can cost more than a billion US dollars on an average.[25] This, in turn, means higher drug prices and consequently, poorer access to healthcare. AI has the potential to change this scenario dramatically.

The International Cancer Genome Consortium aims to profile about 350,000 cancer patients and have already finished profiling 20,000 of them. This would mean having real time data of about 350,000 cancer patients; a goldmine for further medical research. With increased computational capability and data storage capability, these data could be analyzed and assessments reached quickly. Coupled with this, the Consortium has also developed CANSAR database, an integrated data base, which collects genetic data from previously conducted studies and the effects of certain drugs. It uses AI to predict risk of a person becoming cancer ridden.[26]

Apart from research, AI can be helpful in actual treatment procedures such as microsurgery. This could be done with more precision with the help of AI than direct human intervention. This is especially true where the organ on which the surgery is to be done is too small in size and may cause a lack of precision. Further, AI powered chatbots (i.e., chat robots) are also being contemplated.[27] The chatbots will ask a set of multiple-choice type of questions to the patient. Based on these data, a remedy could be recommended by the AI chatbot. According to Juniper Research, chatbots are going to save US$ 8 billion worth of costs in the healthcare sector by 2022.[28]

C. Space Exploration

AI is already in use when it comes to space exploration. For example, if an exploratory object such as Mars Rover had not used AI systems but was remote controlled, it would not have been able to do what it did. In the absence of AI, it would take about 15-20 minutes for signals from Earth to travel to Mars.

So if the object on Mars was moving towards say the edge of a cliff, the command from Earth to stop itself would reach it only after it had already fallen off the cliff. This is why the Mars Rover employed AI to help its movements.

AI has played a crucial role in the most recent space exploratory developments as well. The Event Horizon Telescope (EHT), using AI algorithm, reconstructed the image of a black hole. Without the algorithm capturing such an image would have required a telescope that would be the size of Earth itself.[29] The EHT is an international collaboration that had telescopes around the globe. These telescopes took partial images of the black hole and hence had some gaps. These were filled by the AI algorithm when it reconstructed the full image.

V. AI and its Limitations

Like any technology, AI has its own limitations. While there are many limitations, attention here is focused on three: Discriminatory Behavior, Lack of Accountability and Inability to Introspect.

A. Discriminatory Behavior in AI

As explained earlier, Machine Learning essentially involves analysis of data through an algorithm fed into the Machine. On the other hand, Deep Learning is strikingly similar to the manner in which humans learn, by way of neural networks, albeit artificial ones.

The purpose of AI is to make machines more human-like and the way to make machines similar to humans is to enable them to self-learn, in essence, Deep Learn. However, while it is easy to understand where a Machine Learning algorithm went wrong and why it did not reach a desired conclusion, the same is not possible for artificial neural networks and herein lies a major problem. Illustratively, if a machine that has learnt by way of Deep Learning reaches a conclusion that a particular race of humans is superior to all other races, or a certain sex is superior to another sex, it will not be possible to pin down where that machine went wrong. This is, of course, not to suggest that AI, which is a result of Machine Learning algorithms and not neural networks do not run the same risk. The only difference is this: while in the case of algorithm-based machine learning it can be ascertained where the things have gone wrong, in the case of artificial neural network-based Deep Learning, the same is not possible.

These apprehensions are valid. There have been situations when AI has given undesirable results. One such case may have occurred in the United States where courts use AI known as COMPAS (in cases related to parole, bail or sentencing) to ascertain the risk in granting such parole, bail or sentencing. The assessment made by Compas is taken into consideration by the Judge while passing the final order. In the case of *Wisconsin v. Loomis,* the accused was given a long sentence in consideration of suggestions made by Compas. The said order was challenged on the grounds that the accused was not allowed to assess as to why Compas made such a suggestion, and that such a system lacked transparency. The Supreme Court ruled and disallowed the assessment of Compas. In 2016, the news organization *ProPublica* said that Compass was racially biased.[30]

A similar situation arose when Amazon realized that the software it used to recruit its employees was behaving in a discriminatory manner. In 2014, the company was working on developing AI that would choose the best amongst a number of resumes received by it. The search engine was trained to observe patterns in resumes over a ten-year period. Since the tech industry is dominated by men and more men applied for tech jobs than women, it assumed that men are better candidates just by virtue of being men.[31]

B. Lack of Accountability

It is nearly impossible to understand why Deep Learning algorithms reach a certain conclusion. This is critical when it comes to accountability. When one consults a doctor, he or she gives us a prescription. One can ask the doctor what all factors were taken into consideration before he or she gave the prescription. This possibility of holding the practitioner accountable gives a sense of trust. On the other hand, the lack of accountability of AI may be the reason why people may not yet trust AI with their bodies. There are also privacy concerns with regards to patients' medical records that would be used as data for training AI.[32]

C. Inability to Introspect

It must be recognized that as the machine becomes more and more like the human brain, the biases that humans adopt may also appear in them. The only difference would be that humans have the innate ability to introspect and realize when they are being biased while the machines do not, at least as yet.

Humans develop biases by way of noticing patterns around them (inductive logic). For example, if larger number of women are involved in road accidents than men, then the human mind may notice this pattern (typically from a small sample) and come to the erroneous conclusion that women are bad drivers. However, humans also have the ability to understand when their conclusions may be biased. A machine which has trained itself by way of Deep Learning also learns in a similar manner, but it does not realize when it becomes biased.[33] More importantly, given the inability to introspect, it is not possible to determine why and where a machine with artificial neural networks has reached a certain biased conclusion.

Notably, it has been argued that human societies, over a long period of time, have evolved the moral rule that killing or injuring another person is unacceptable in most circumstances.

Therefore, when a person grows up in a society, he or she almost inherently adopts this rule, and this is why it is important that a decision of killing or injuring should always be a human decision. Only a human can understand the moral guilt of killing someone innocent and consequently, he or she is likely to take more careful decisions. Deep Learning machines, on the other hand, neither provide the possibility of a human assessment of where it went wrong and why it came to an undesirable conclusion (which in this case could be as serious as killing an innocent human being), nor has ability of moral introspection to realize where it may have gone wrong.

Such a shortcoming of AI, i.e., the lack of moral introspection may be the Achilles heel of either the development of AI or the human civilization itself.[34]

VI. AI Control Problem

AI control problem has been subject of several sci-fi movies and books. The typical story of Robots rising to cause extinction of humans has become almost a cliché. However, this may not be a distant nightmare.

Most people are familiar with Science Fiction stories by Isaac Asimov. His stories in three volumes revolve around violation of Asimov's Three Laws of Robotics:[35]

i.) First Law: A Robot may not injure a human being or through inaction, allow a human being to come to harm.

ii.) Second Law: A Robot must obey orders given to it by human beings except where such orders would conflict with the First Law.

iii.) Third Law: A Robot must protect its own existence as long as such protection does not conflict with the First or Second Law.

It is through these laws (or algorithms) that the humanity was supposed to be protected from a take-over by Robots.

As AI technologies progress, the frontier moves further and technologies that are called AI today lose the status of being AI. In this process, people become more and more dependent on AI. When humans are on such a path they must ask themselves where they are headed. The problem of AI control raises precisely this question.[36]

The easy answer to such concerns is to focus on making AI that is conducive to human goals and aspirations. However, this is easier said than done. Since the problem is not that simple, the solution also cannot be a straightforward one.

What makes the AI Control Problem far more serious than what appeared earlier is the fact that machines becoming more intelligent than humans in the future, now seems to be inevitable. This is simply because improvements in existing technologies will not stop. There is no compelling reason why the propensity of human beings to constantly improve technology, manifest itself throughout human history, will suddenly come to a grinding halt.

As AI technology develops, a point would come when an ultra-intelligent machine will surpass human intelligence and as I.J. Good, the Cryptographer who worked alongside Alan Turing said, an intelligence explosion will occur.[37] This would mean that machines would then be capable of making technological leaps, within a short span of time which would have taken human beings thousands of years. According to Good, these machines will create newer machines which would be technologically more advanced, and this "explosion" in AI would leave human intelligence far behind.

In order to appreciate how real and impending this future is, consider this: the existence of machines that improve themselves are not a part of human imagination any longer. There are machines that employ Deep Learning in order to improve themselves at tasks handed to them. Self-programming machines are already a reality and in contrast to humans, the rate at which they improve is exponential. Even if AI does not surpass human intelligence, and remains only as intelligent as human beings, in view of the fact that electronic circuits function several times faster than biochemical circuits, such AI would still be several times faster than an exceptionally intelligent human mind. It would be able to make thousands of years' worth of human intellectual advancements within weeks.[38]

The impact of global geo-politics is also an important factor to consider while talking about controlling AI. In a world where arms race amongst nations is a common phenomenon, along with economic competition for ongoing opportunistic acquisition of power, it is obvious that countries will try and acquire better AI technologies sooner than their competitors.

This process has already begun; the UK, USA, Japan, China and India have already come out with strategy papers for promotion—not regulation, of AI research. This is, to some extent, understandable because in this race whoever starts first would gain a great advantage over others.

Eminent technology geniuses such as Elon Musk who believe that AI is a greater risk to human existence than nuclear bombs, and that it could be an immortal dictator, suggest regulation of AI research as a solution. However, as mentioned before, nation states and corporates as of now are more concerned with winning the AI race, than keeping the genie in the bottle.[39]

The issues relating to regulation and strategic development of AI and other revolutionary technologies have been taken up for detailed discussion in Chapter 13.

Annexure I
Pythagorean Theorem Proof[40]

The Pythagorean Theory says: $a^2 + b^2 = c^2$ in the figure below:

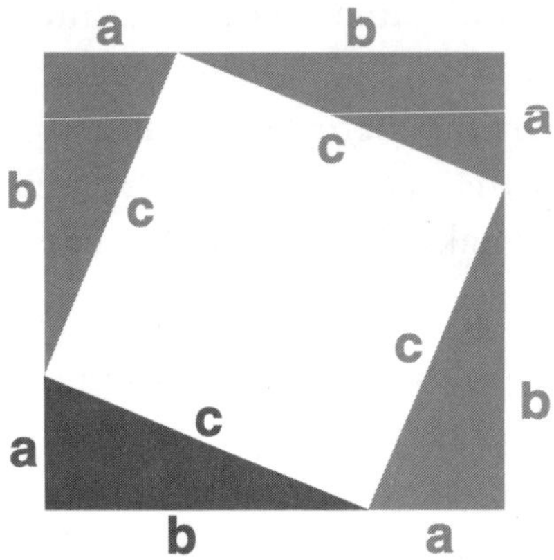

The area of the large exterior square can be expressed as $(a+b)^2$ (Premise 1) or same may also be expressed by summing up the area of the four triangles and the area of the smaller interior square i.e. $4(1/2ab)+c^2$ (Premise 2).

Since both of them would separately calculate the area of the same large square, then both of them must be equal. (Premise 3) Hence, we can say that $(a+b)^2=4(1/2ab)+c^2$ (conclusion).

i.e., $a^2+2ab+b^2=2ab+c^2$ (Simplified version of the above conclusion)

Hence, it follows that $a^2+b^2=c^2$.

3

The Future is All About Data

"It is a capital mistake to theorize before one has data."

Sherlock Holmes[1]

Introduction

It was Clive Humby, a British Mathematician who reportedly said it first in the year 2006, "Data is the new oil. It is valuable, but if unrefined it cannot really be used". Not only the oil, but this statement itself has been refined over the years. In 2013, Abhishek Mehta, the then CEO of Tresata said, "Just like oil was a natural resource powering the last industrial revolution, data is going to be the natural resource for this industrial revolution". The most interesting further refinement came from Piero Scaruffi, a cognitive scientist (and author of *History of Silicon Valley*) in the year 2016. He said, "The difference between oil and data is that the product of oil does not generate more oil (unfortunately), whereas the product of data (i.e. from self-driving cars, drones,... etc) will generate more data (where do you normally drive, how fast/well you drive, who is with you, etc)."

This chapter is all about Big Data and the future is all about Big Data. The chapter has been organized as follows:

Section I explains the basics: what is Big Data and Data Analytics. Section II brings out the role of Big Data and Data Analytics. Demonstrating the nexus between Big Data Analytics and Artificial Intelligence (AI), it is explained how together they drive the emerging New-Age Technology. Section III then discusses the all-important phenomenon of cross-border data flows. Finally, Section IV delves into the pros and cons of the so-called "Digital Globalization".

I. Big Data and Data Analytics: The Basics

Clifford Stoll, an American author summarized the general understanding about data very aptly when he said, "Data is not information, information is not knowledge, knowledge is not understanding, (and) understanding is not wisdom."

Be that as it may, it is also true that "without data, you are just another person with an opinion", said W. Edwards Deming. In fact, Geoffrey Moore has gone further to say that "Without Big Data, you are blind and deaf and in the middle of a free way."

Admittedly, some skeptics say that "Torture the data long enough, and it will confess to anything" (Economist Ronald Coase and Econometrician Edward Leamer). Nevertheless as pointed out by a Swedish mathematician Andrejs Dunkels, "It is easy to lie with Statistics. It's hard to tell the truth without statistics."

Most people would agree with what Daniel Keys Moran said, "You can have data without information, but you cannot have information without data." To hark back to Napoleon Bonaparte, "War is ninety percent information".

It all boils down to the challenge of converting data into information and then transforming that information into operationally meaningful insights.

And, that is precisely what Big Data and Data Analytics do.

There is a long tradition of analyzing large datasets to develop insights. These insights can be fundamental to how a business operates and expands. For instance, consider an airline that wished to open a new route from point A to point B. It would have to check the number of passengers commuting through the route using train or any other common mode of travel, and the prices they were willing to pay. It would need huge datasets from the Railways and Bus Services to justify whether or not servicing the route by air would be a good idea. The datasets in this scenario are "Big Data," and the practice of deriving insights therefrom is "Big Data Analytics."[2]

Big Data typically includes datasets with volumes or sizes that are "beyond the ability of commonly used software to capture, curate, manage and process data within a tolerable elapsed time".[3] Managing and processing Big Data involves storage, sharing, transfer, updating and of course, analysis of data.

How big should be the data size so as to qualify as Big Data? It is actually a moving target or rather ever-increasing moving target. This is so because the world's technological capacity to store, communicate and compute information has been doubling every 40 months since the 1980s.[4]

Big Data has been expanding extremely rapidly, indeed. An estimate in 2013 showed that 90 per cent of all the data in the world had been created in just two years preceding 2013.[5] *Forbes* recently reported some mind-blowing statistics: as much as 2.5 exabytes of data is produced every day.[6] [1 exabyte = 1 billion gigabytes and is equivalent to 2 million Personal Computers (PCs)].

More formally, Big Data has been defined as "the information assets characterized by such a high volume, velocity and variety to require specific technology and analytical methods for its transformation into value".[7] In these characteristics of Big Data, "volume" refers to the quantity of generated and stored data; "variety" implies the type and nature of the data, such as text, images, audio, video, etc, whereas "velocity" means frequency of generation and processing of the data. Some experts add another V to the characteristics of Big Data – Veracity: accuracy or quality of data. The basic premise here is that a combination of these four Vs—volume, velocity, variety, and veracity—that creates the fifth V which is value for the Big Data. The main instrumentality of achieving the fifth V i.e. value, is Data Analytics.

II. Role of Big Data and Data Analytics

Big Data and Data Analytics is a fast-growing industry. In 2010 *The Economist* had estimated that the industry was worth more than US$ 100 billion and was growing at around 10 per cent per year, about twice as fast as the whole software business was expanding then.[8] Going by that prediction, the industry today could be worth around US$ 250 billion.

The use and adoption of Big Data and Data Analytics, whether in government or in the private sector, can promote cost-efficiency, productivity and innovation.

At the macroeconomic level, Big Data and Data Analytics offer cost effective opportunities to improve decision-making in major development areas such as overall productivity, resource management, education and health care, besides crime, security and natural disaster management.[9, 10]

At the micro level, Big Data Analytics can help businesses expand through streamlining three areas: research and development, decision making as well as customer satisfaction. Big Data Analytics enables companies to identify patterns of demand (trend mapping) and larger business opportunities. Not only do these insights provide scope for faster advancements of products, but could also allow companies to respond to changes in the business environment in real time or even beforehand. Illustratively, BMW – the German automobile giant, recently leveraged this technology to their advantage by creating an "idea management system" to evaluate ideas submitted through its "virtual innovation agency". Collecting this data reduced the company's time spent on identifying high-potential ideas by 50 per cent and led it to annually incorporate two to three ideas from the open innovation effort into new car models.[11]

Big Data Analytics has also been helpful in guiding decisions in our day-to-day life. Data generated and transferred across borders has become a convenient tool simplifying the daily life today. For example, as most people are aware, collecting and analyzing locations through mobile phones allows Google Maps to help determine the best (if not the shortest) path from point A to point B. Insights developed also help users to remotely check for traffic congestion as well as estimate the amount of time it might take to reach their destination.

Insights from Big Data Analytics thus, have already been driving innovations and economic growth. Needless to say that as Big Data gets bigger at an unprecedented pace, more trends and greater operationally meaningful insights are bound to emerge. In fact, it is already happening.

The confluence of the emerging technologies, especially Cloud Mobility, Internet of Things (IoT) and Data Analytics have greatly amplified the data explosion, thereby leading to incredible growth of un-structured data. About 80 per cent of data generated by businesses is unstructured, which is precisely why Big Data Analytics and AI have emerged as strategic assets.[12]

Big Data empowers AI and plays a critical role in its development. In simple terms, Data is really the fuel that AI runs on. AI algorithms analyze huge sets of database to uncover patterns that would be almost impossible to identify with human eyes. The more data AI has, the faster and better it can learn from it. In effect, AI and Big Data share a mutually reinforcing relationship. AI algorithms can learn and develop through Big Data and derive insights that may not be attainable through just human intelligence. Moreover, it would require only a fraction of the cost needed to conduct such analyses manually. In turn, Big

Data grows through these patterns, possibly identifying new areas of data collection, leading to more insightful problem solving. Thus, when it comes to Big Data, more is truly merrier for AI.

Notably, not only does the quantity of Big Data matter, the scope and quality are also important. For example, by using just the Indian data, AI may solve problems which are India-centric. But using say, South Asian data, AI may solve problems at a regional level as well. So more diverse the data sources are, the better it is for AI.

What does the free flow of Big Data portend for an AI program that feeds on data itself? It would mean the whole world. There are two reasons why: first, it would increase the sheer volume of learning material. Secondly, it would mean variety in the scope of data available. Both of these could mean greater accuracy in drawing inferences and decision making, thus adding more value.

Let us consider the former first. If AI is given data for only one country to process, it will develop insights, learn from the patterns it observes in that country and solve the problems identified. However, in this scenario AI would not be able to realize its full potential. AI would learn a whole lot more if it is presented the data for 10 countries instead of just one. This would also invariably broaden the range of the problems that AI can tackle.

This discussion logically leads to the all-important phenomenon of cross-border data flows, which is taken up next.

III. Cross-Border Data Flows

As discussed earlier, the emerging New-Age Technology has been transforming the way people interact, communicate, exchange and share.

Millions of individuals and businesses communicate, buy and sell *via* the Internet every day. Every audio-call on Internet, Skype-call, e-mail, social media post, Internet search and each online sale-purchase transaction involves transfer of information or data from point A to point B (i.e. data-flows). When placed in a global context, data flows that cross national borders are cross-border data flows. Notably, data flows are no longer confined to the Internet. In fact, as explained in Chapter 4, the Internet of Things (IoT) provides a fast growing mechanism to connect "smart" physical objects, such as TVs, home appliances, cars, etc.

Cross-border data flows are virtually exploding. According to one estimate, by 2020:

i.) there will be global web-traffic (i.e. the flow of data across the Internet) amounting to 194 exabyte per month. (It may be recalled that an exabyte equals one billion gigabytes.)

ii.) the number of global Internet users would shoot up to around 4.1 billion (i.e. more than 52 per cent of the world's population).[13]

The McKinsey Global Institute (MGI) Report (2016) estimated that "cross-border bandwidth usage" increased 45-fold since 2005 and is projected to grow another nine-fold by 2020".[14]

According to Lund and Manyika (2017) "as global trade in goods and services has plateaued and global financial flows have declined, dramatically, cross-border data flows are exploding in volume."[15]

Assessing how important cross-border data flows might be for the global economy is an exceedingly difficult task. A World Bank study, using data from 2001 to 2013, attempted to measure the impact of Internet penetration on bilateral trade. The study concluded that a 10 per cent rise in Internet penetration in the exporting country brings about a 1.9 per cent increase in the value of exports.[16] Moreover, there are additional ripple effects of cross-border data flows on different economies in whole sectors (such as supply-chain management) which are difficult to quantify.

Measuring the (economic) value of cross-border data flows is a complex matter.[17] However, the following statistics may be taken as indicators of its enormity:

1. Cross-border data flows have already been making a larger contribution to global GDP than trade in goods. Between 2004 and 2014 global flows of goods, services, people, finance and data together raised the global GDP by at least 10 per cent. In 2014 alone, these factors collectively added US$ 7.8 trillion to the world economy. "Remarkably, cross-border data flows – which were negligible just 15 years ago – accounted for US$ 2.8 trillion of this value (i.e. 36 per cent), surpassing the impact of the global goods trade, which has evolved over centuries."[18]

2. Global e-Commerce accounted for roughly 10 per cent of the world global trade in goods (value) in 2015. In China, nearly 20 per cent of exports and imports are routed through digital platforms, which is about twice the share of e-Commerce in Europe.[19]

3. The estimated value of global business-to-business e-Commerce in 2013 was over US$ 15 trillion of which, 75 per cent was accounted for by the US, the UK and China, in that order.[20]

4. Yet another measure relates to the so-called "potentially ICT-enabled services" i.e. "services that can predominantly be delivered remotely over ICT networks" and "include activities that can be specified, performed, delivered, evaluated and consumed electronically."[21] In 2014, 54 per cent of total US services exports and 48 per cent of total US services imports were accounted for by "potentially ICT-enabled services". For the 28 member states of the European Union, these exports amounted in aggregate to US$ 1.2 trillion in 2014, emanating from the UK (US$ 159 billion), Germany (US$ 149 billion) and France (US$ 115 billion).[22]

5. According to an estimate, roughly 12 per cent of global goods trade is conducted *via* international e-Commerce. In addition, 361 million individuals around the world participate in cross-border e-Commerce.[23]

The economic impact of cross-border data flows is thus, highly significant. It has opened up avenues for international trade to new market players and places. For example, online market places such as Amazon and Alibaba are hosting millions of small merchants from all over the world. The number of small companies on Facebook shot up from 25 million in 2013 to over 50 million in 2016.[24]

The availability of cross-border data flows helps AI in a way that can transform it to solve the problems on a broader scale. Possession and processing of these datasets could facilitate regional or even global decisions. The more the global data AI has, the larger would be its scope and scale. In fact, AI would not be confined to observe patterns only at a national level, but on a regional or global one and therefore, its efficacy could be greatly enhanced.

As information technology penetrates deeper into human activity, and more data is generated by people, businesses and increasingly from sensors embedded in products (i.e. the Internet of Things), this activity is only going to grow substantially for the foreseeable future.

In sum, cross-border data flows can elevate AI to solve problems on a broader level, speed up learning and may help it find complementary and/or supplementary patterns in regional or even global demand.

Scaling under the Industry 4.0 is cheaper than ever before. Illustratively, global market leader in logistics-DHL has estimated that (Big Data Analytics and) IoT technologies will have an impact of more than US$ 1.9 trillion on supply chain management and logistics.[25]

Unlike the earlier Industrial Revolutions, the impact of technological changes underlying the Industry 4.0 are felt by individuals as well as businesses at unprecedented speeds. However, it may be noted that scaling and automating at the prevailing affordable rates has been made possible because of the reduced costs at the backend. Cross-border data flows are essential for keeping these costs down while simultaneously contributing to innovation and efficiency.

IV. "Digital Globalization" Controversy

The spectacular increase in both the size and the value of cross-border data flows has prompted coining of a new expression: "Digital Globalization".

Of late, "Digital Globalization" is being challenged in several countries with imposition of a number of barriers and putting in place various protectionist policies. For example, "the set of Internet restrictions known as the Great Firewall of China", and Russia's insistence on data storages (containing data on its citizens) to be located within. Several countries have banned instant messaging services like WhatsApp and have blocked selected websites. Protectionist policies also may include online censorship, market restrictions on digital content providers and conflicting or overlapping rules on data privacy and protection.[26]

More often than not, the reasons for such digital protectionism include entirely legitimate concerns about data privacy, cybersecurity and "protecting domestic digital platforms" from foreign competition. While the protectionist policies may be well intentioned, they invariably threaten to disrupt the free flow of cross-border data, which entail significant costs to economies and harm to consumers.

Multiple arguments have been put forth in defence of the "Digital Globalization."

i.) Instead of promoting cross-border data flows, if they are restricted for whatever reasons, arguably, it "would mean missing out on the regional scope that AI has to offer. In addition, it would also serve as a hurdle that would hinder the potential that AI might develop in the future".[27]

ii.) The importance of sharing data cannot be overemphasized. In its significance, it is perhaps second only to the generation of data itself. Big Data can be shared at unprecedented speeds and across national borders. As a result of high-speed cross-border data flows, Multinational Companies (MNCs) with Internet of Things (IoT) technology have been able to optimize their services. Cloud computing services is a good example. These MNCs can provide inexpensive, on-demand computing capability to big and small businesses alike. Accordingly, small and medium firms in developing countries can avail world-class Cloud computing services as per the computing power they need. In this process, entire sectors such as supply chain management are being transformed across the world.

Notwithstanding this controversy surrounding "Digital Globalization", there is no doubt that genuine concerns against the free cross-border flows must be addressed, possibly through the multi-lateral framework of global co-operation, rather than restrictive protectionism as emphasized by the latest G-20 Declaration.[28] This would ensure that the world economy does not miss out on the tremendous potential benefits of Digital Globalization.

4

Robotics, Blockchain, Automated Technology, Additive Manufacturing, Internet of Things and Augmented Reality

"Man is to technology what the bee is to the flower. It is man's intervention that allows technology to expand and evolve itself and in return, technology offers man convenience, wealth and the lessening burden of physical labor via its automated systems".

James Scott Brown[1]

Introduction

As noted earlier, New-Age Technology underlying the Industrial Revolution 4.0 is really a post-Digital Revolution Fusion Technology. In actuality, it is a fusion of several different technologies such as Artificial Intelligence (AI) and Machine Learning (ML), Big Data Analytics, Robotics, Blockchain Technology, Automated Technology (AT), Additive Manufacturing (AM) and 3D Printing, Internet of Things (IoT) and Augmented Reality (AR), etc. Having covered AI and ML (Chapter 2) and Big Data Analytics (Chapter 3), this Chapter is dedicated to the remaining prominent constituents of the emerging revolutionary technology.

From Robotics to Internet of Things (IoT) and Augmented Reality (AR), this chapter explores the realms of technology advancements that are slowly, but surely, playing a major role in making human lives easier and connected. The fusion of all these technologies enables work processes move faster, enhances productivity and improves quality and safety of human lives.

The chapter has been organized as follows:

Section I discusses Robotics, followed by Blockchain Technology in Section II. Section III covers Automated Technology while Section IV dilates on Additive Manufacturing and 3D Printing. Against that backdrop, Section V covers the basics of Internet of Things (IoT) whereas the final section , i.e. Section VI delves into the technology of Augmented Reality.

I. Robotics

Robotics technology has developed considerably since Isaac Asimov, coined the term in the early 1940s in one of his science fiction stories. It is defined as *"the science or study of the technology associated with the design, fabrication, theory, and application of robots"*. In turn, there are a number of definitions in use for 'Robot'. A reasonable one that is attributed to the Robotics Institute of America (RIA) defines a Robot as a *"reprogrammable multifunctional manipulator designed to move materials, parts, tools, or specialized devices through variable programmed motions for the performance of a variety of tasks."*

The arrival of Robotics technology has gone beyond science fiction, where robots were presented as the crowning glory of mankind's technological success. Since most people pictured robots the way they were presented in science fiction movies, they often expected Robots to be human-like in appearance. However, what a Robot looks like is actually related to the tasks it is supposed to perform. Machines that do not look like humans can also be classified as Robots. On the other hand, some human-looking "robots" are not much beyond mechanical devises or toys.[2]

The confluence of several advanced technologies is bringing the age of Robotics ever near, with emergence of Robots which are smaller, cheaper, more practical and cost-effective.[3] For any Robot, there are three aspects that are important. First, "*brawn*

– strength relating to physical payload that a Robot can handle." Then comes the "*bone* – the physical structure which is relative to the work it does that in turn determines the size and weight of the robot in relation to its physical payload." And finally, the "*brain* – robotic intelligence; what it can think and do independently and how much manual interaction is required."[4]

Robots have evolved tremendously from being big machines with significant brawn but little else in the early days. In the early stage, they were hydraulic-powered and were relegated to dirty and dangerous tasks that humans did not want to touch for various reasons. With the technological advances, Robotics has come a long way. Over the years as things evolved, productivity and strategic benefits of Robots became more evident. For example, by the 1980s, Robots had transitioned from being hydraulically-powered to become electrically driven units.[5]

Today, there are, reportedly, close to a million robots in the world with Japan leading the way with almost half the number, followed by US that deploys 15 per cent of the total number. Robots have been very prominent in the automobile sector. There was a time when nearly 90 per cent of them were deployed in car manufacturing, generally on assembly lines performing a variety of repetitive tasks. This is not the case anymore. That number has now come down to 50 per cent in automobile plants, while the other half is deployed in other factories, laboratories, warehouses, energy plants, hospitals, and so on.

With the emergence of the Industrial Revolution 4.0, Robots will continue to be deployed, for assembling products, handling hazardous materials, spray-painting, cutting and polishing inspection of products. At the same time, they will continue to be used for tasks as diverse as cleaning sewers and detecting bombs.

Over time, Robots have become more intelligent and have begun executing more complex tasks. They are now used as space explorers, for surgeries and even as pets. For example,

Sony has developed a Pet Robot called AIBO, the mechanical dog. It lies down and flips over to play. If you give him a voice command, such as "bring me the phone", it will respond and duly comply with it.[6]

With the emergence of AI and increasing computing power, the potential of Robots to support humans in their tasks has risen manifold. The new surge of robotics is in applications demanding advanced intelligence. "Robotic technology is converging with a wide variety of complementary technologies – machine vision, force sensing (touch), speech recognition and advanced mechanics".[7] This has opened up exciting new levels of functionality for jobs that were never considered feasible earlier for Robots.

In the developed countries, Robots are being used in multiple places. From ATMs to petrol pumps, machines are simplifying and making the tasks easier to perform. The future is here and one cannot ignore it.

At the same time, people also cannot overlook the warning given by the legendary scientist Stephen Hawking: "Unless mankind redesigns itself by changing our DNA through altering our genetic makeup, computer-generated Robots will take over our world."[8]

II. Blockchain Technology

Blockchain is essentially a digital ledger of financial transactions that is non-corruptible. It has been hailed to be "nothing less than the second era of the Internet, with the potential to fundamentally change the modern digital economy".[9]

In its simplest form, Blockchain is just a chain of blocks which contains records and are linked using cryptography. Each block contains timestamp and transaction.[10] "Each transaction is digitally signed to ensure its authenticity and that no one tampers with it, so the ledger itself and the existing transactions within it are assumed to be of high integrity".[11]

A. How Does it Work?

When a new transaction is initiated, a majority of nodes within the Blockchain execute algorithms to verify the history of the individual block. If the majority of nodes come to the conclusion that the history and signature is correct, the new block of transaction is accepted and is added to the transaction. The distributed consensus model allows the Blockchain to run (i.e. carry out financial transactions) without the need of central banks or an authority validating the transaction.[12]

The step by step process of this technology is as under:[13]

1. A node starts a transaction by first creating and then digitally signing it with its private key (created via cryptography).
2. A transaction is propagated (flooded) by using a flooding protocol, called Gossip protocol, to peers that validate the transaction based on preset criteria. Usually, more than one node is required to verify the transaction.
3. Once the transaction is validated, it is included in a block, which is then propagated onto the network. At this point, the transaction is considered confirmed.
4. The newly-created block now becomes part of the ledger, and the next block links itself cryptographically back to this block. This link is critical. At this stage, the transaction gets its second confirmation and the block gets its first confirmation.
5. Transactions are then reconfirmed every time a new block is created. Usually, six confirmations in the network are required to consider the transaction final.

The biggest example of use of Blockchain Technology is Bitcoin, which employs public ledger in which anyone can participate.

B. Features of Blockchain Technology[14]

1. **Peer-to-Peer**: There is no centralized authority to control or manipulate. All participants communicate with each other directly. This enables data exchange directly without any third-party participation.
2. **Distributed**: The ledger is spread across the whole network, thereby making tampering very difficult.
3. **Cryptographically Secured:** Cryptography is used so as to make the ledger tamper-proof.
4. **Add-Only:** Data can be added in Blockchain only in a time-sequential order. This implies that once data is added to the Blockchain, it is almost impossible to change it.
5. **Consensus**: Blockchain Technology has the ability to update the ledger *via* consensus. This gives it the power of decentralization. There is no central authority in control of updating the ledger. Any update made to Blockchain is approved using well-defined and strict criteria specified by the Blockchain protocol and data is added to the Blockchain only after a consensus has been reached among all participating peers/nodes on the network.

Blockchain Technology has been regarded by some as the future of everything.[15]

III. Automated Technology

Automation is the execution of tasks without periodic interference that minimizes and gradually does away with human intervention.

A software operating on its own and an electronic device functioning autonomously are some examples of automation.

Automation simplifies complicated multiple tasks by reducing them to a single one. For example, a press of a button or a simple command will set into motion a chain of events towards a specified objective. Today, automation is widely deployed across industries. Automatic devices monitor the necessary steps on their own, leading to cost optimization and timeline management. This also facilitates identification of problems and their redressal so as to ensure seamless working. As Safra Catz puts it succinctly, "a true digital transformation takes advantage of technology to focus on the customer, automates work that does not need manual interference, and unleashes your people to truly make decisions that change the path of your company".[16]

Automation is used in many sectors to do multiple tasks and enhance efficiency. For example, automation technology is applied in customer-interaction solutions. This includes kiosks, interactive digital signage, and self-checkout counters. Improved by recent innovations (better touchscreens, faster processors, improved connectivity and sensors), it is also entering new markets and applications, a prime example being the experimental Amazon Go convenience store.[17]

Automation is being heavily used in factories across the world and that has brought tremendous gains to the industry. There are multiple advantages of automation, such as higher productivity, better production rates, enhanced efficiency and safety.

By "automating" certain tasks carried out in factories and various work places, enterprises are able to make better and more efficient use of its labor force. Moreover, automated systems typically perform the manufacturing process with less variability than human workers, resulting in better control and improved consistency of product quality. Also, increased process control makes more efficient use of materials, resulting in less scrap.[18]

It is often argued that one of the disadvantages with automation is high capital expenditure required to place the systems. Automation is still a relatively new technology that will require some more time to reach scale that would bring down the initial cost of investing. While there is an element of truth here, the argument is fast becoming obsolete. For example, as pointed out by Anthony Stieber, "Any daily work task that takes 5 minutes will cost over 20 hours a year, or over half of a work week. Even if it takes 20 hours to automate that daily 5 minute task, the automation will break even in a year".[19]

Another major disadvantage of automation is worker displacement. This is a hard reality with every generational change or shift in technological patterns. The answer to meet this challenge is to gear up the labor force with the latest skills. Later in the book (i.e. Chapter 5), this challenge has been analyzed in detail.

Here again, the warning given by Stuart Chase, does have a certain intuitive appeal:

> Labor-saving devices, led by the computer, are encouraging more production and more unemployment. Automation may thus be classified as both asset and liability. It can eliminate a great deal of dull, monotonous labor, but if unrelieved idleness is the end result, the cure may be worse than the disease. The evidence strongly indicates that man is a working animal—in the sense of needing an activity that involves hand and brain, and that makes sense to him.[20]

IV. Additive Manufacturing and 3D Printing

Additive Manufacturing technologies build 3D objects by *adding* layer-upon-layer of material, may it be plastic, metal, concrete or even human tissue. It encompasses a range of technologies such as 3D Printing, Rapid Prototyping (RP), Direct Digital

Manufacturing (DDM), layered manufacturing and additive fabrication. "It is being used to fabricate end-use products in aircraft, dental restorations, medical implants, automobiles, and even fashion products."[21] It improves the production of highly complex tools and components. Indeed, it makes their manufacture possible in the first place.

A related concept is that of 3D Printing, which is the "action or process of making a physical object from a three-dimensional digital model, typically by laying down many thin layers of a material in succession".[22]

Interestingly, 3D printing can basically produce pretty much anything from cups to toys, to metal and machine parts, spectacles, cakes and furniture. The list is endless. Gradually, the technology is moving into an arena where it may be able to print even human body parts. 3D printing is able to replace factory production lines with a single machine, just like home inkjet printers replaced bottles of ink, a printing press, hot metal type and a drying rack.[23]

To be sure, this unprecedented capability has a dark side as well. Additive Manufacturing will make it easier for individuals to manufacture weapons. The technology, therefore, involves serious security risks. It is worth recalling what Professor Neil Jacobstein had reportedly said about being worried not about artificial intelligence, but about the human stupidity.[24] This dimension of weaponization of Additive Manufacturing and 3D Printing has been taken up later in Chapter 12 of the book.

V. Internet of Things

According to O'Brien Brendan, "If you think that the Internet has changed your life, think again. Internet of Things (IoT) is about to change it all over again."[25] Internet of Things (IoT) is all about the power of the Internet being extended beyond just computers and smartphones, to a whole range of processors and external environment. The Internet of Things is a network, truly a technology-enabled "Network of all Networks".

IoT is essentially an ecosystem of connected devices, which uses powerful processors and sensors to receive and send data they acquire from the environments they are placed in. The IoT web consists of an IoT gateway (where all information collected by the sensors is sent) that is connected to the Cloud. This data is then analyzed and acted upon by the devices. Many a times, the IoT devices communicate with each other and act on information they receive and send. "The connectivity, networking and communication protocols used with these web-enabled devices largely depend on the specific IoT applications deployed."[26]

IoT web helps people with tasks and learn from processes. It allows businesses and people to stay more connected, which in turn, helps them coordinate better, thereby enhancing efficiency, cutting down on waste and enhancing productivity of their work.[27]

When something is connected to the Internet, it means that it can send or receive information, or both. This ability makes things 'smart'. To be smart, a thing doesn't need to have super storage or a super-computer inside it. All that it has to do is to connect to a super storage or a super computer.

According to Grant Notman, "The Internet of Things is removing mundane repetitive tasks or creating things that just weren't possible before, enabling more people to do more rewarding tasks and leaving the machines to do the repetitive jobs."[28]

IoT has multiple uses in the real world ranging from in consumer applications to enterprises and manufacturing. IoT applications cover several sectors, including automotive, telco, energy, and more. The emergence of smart appliances heralds the start of the IoT era. This has led to enhanced levels of efficiency that couldn't have been imagined before. For example, consider the growth of "smart" homes across the world. The ability to control home or office devices remotely is possible now only because of IoT. The future is right here:

> Suppose one fine day you decide to come back home early from work. In an IoT enabled home, you can simply switch on your geyser so that the water is heated by the time you reach home, the washing machine cleans your clothes and the toaster has made your favourite sandwiches that you activated while in the cab. If you ever forget to switch off lights while leaving for work, you needn't worry. If your home is connected to your phone, with smart lighting enabled, the lights can be turned on while you skip through your mails in the cab. In the future, when the devices are all interconnected, we will witness a more efficient living that will help us save energy, waste less thereby contributing to a healthy living.[29]

Similarly, IoT has multiple benefits in healthcare. With the emergence of the technologies underlying Industrial Revolution 4.0, especially combining the forces of AI and IoT can indeed revolutionize medical ecosystem in the future. Wearable devices that track BP and heart rate are just the starting point of this revolution. These wearable devices collect extensive data about the patient's heart rate, metabolism, breathing rate, blood pressure, etc., and transmit this data to the Cloud, where it is processed and analyzed in different ways and provide cutting edge insights about the human body behavior. Accordingly, doctors can prescribe more accurate medicines, perform the right diagnostics and therefore, enhance the lifeline of patients.

According to Mehul Nayak, "Internet of Things is changing and transforming everything from business to life. Imaginations are boundless and opportunities are infinite. Everything is being wired up or connected wirelessly".[30]

No wonder, a report from PricewaterhouseCoopers has made the following prediction:

> "Internet of Things is transforming the everyday physical objects that surround us into an ecosystem of information that will enrich our lives. From refrigerators to parking spaces to houses, the Internet of Things is bringing more and more things into the digital fold every day, which will likely make the Internet of Things a multi-trillion dollar industry in the near future".[31]

VI. Augmented Reality

According to Tim Cook, CEO of Apple, "a significant portion of the population of developed countries, and eventually all countries, will have Augmented Reality (AR) experiences every day, almost like eating three meals a day. It will become that much a part of you."[32]

Augmented Reality is essentially an enhanced version of reality. Here in, computer-generated images are superimposed on a user's view of this real-world environment. This augmentation enhances one's prevailing perception of reality.[33] Unlike virtual reality, which requires one to go into an entirely virtual environment, augmented reality simply overlays virtual information on top of it the natural environment. Since both, virtual and real worlds harmoniously coexist, users of augmented reality experience a new and improved natural world.[34]

The technology of Augmented Reality is already being used in multiple ways. Google SkyMap for example, is an AR app that overlays information about constellations, planets, etc. when one points the camera of the smartphone towards the sky. Wikitude is another app that provides information about a landmark or object by simply pointing a camera of smartphone at it.

As Muchaneta Kapfunde puts it: -

> "This [Retail] is a sector where AR is revolutionising the way we shop by bridging the gap between digital and physical worlds. A radical store re-invention will offer retailers new ways to increase convenience and personalisation, and consumers the opportunity to experience re-envisioned shops".[35]

And, needless to say, there is more where all this is coming from!

PART • TWO

NEW-AGE TECHNOLOGY: POTENTIAL IMPLICATIONS

5

Technology: Economic Growth and Human Development

"The twin revolutions in information technology and bio-technology confront us with the biggest challenges our species have ever encountered. The merger of infotech and biotech might soon push billions of humans out of the job market and undermine both liberty and equality. Big Data algorithms might create digital dictatorships in which all power is concentrated in the hands of a tiny elite while most people suffer not from exploitation, but from something far worse that is irrelevance."

Yual Noah Harari[1]

Introduction

The emerging technological revolution that underlies the Fourth Industrial Revolution is throwing up immense opportunities and possibilities. At the same time, in its wake, the unfolding transformative technology has given rise to a series of complex questions and far-reaching societal concerns.

This chapter addresses probably one of the most important set of these issues i.e. the potential impact of the emerging technology on global economic growth and human development. The chapter has been organized as follows: Section I discusses the positive side of technology-led economic growth. Section II elaborates on potential positive impact of the emerging revolutionary technology on human development. Against that background, Section III delves into the negative side i.e., technology-led joblessness. Having discussed the future of jobs in the preceding section, Section IV covers the jobs of the future.

I. Technology-led Economic Growth

Several studies have reported that Artificial Intelligence (AI) *alone* could significantly increase labor productivity and boost the overall global GDP considerably. This would generate substantial additional wealth for many developed as well as emerging market economies. If AI is combined with the other New-Age Technologies such as Robotics, Blockchain Technology, Automated Technology (AT), Additive Manufacturing (AM) and Internet of Things (IoT), the overall positive impact could be enormous.

According to a Report by McKinsey Global Institute,[2] AI alone would contribute, on an average, 1.2 per cent per year in productivity growth. Given the 2018 level of global GDP (at current prices) of US$ 85 trillion,[3] this would mean an incremental addition to global Gross Domestic Product (GDP) of more than US$ 1 trillion per year, thus amounting cumulatively to around US$ 13 trillion, by the year 2030. Just to appreciate how large this additional GDP is, consider this: it would be roughly equivalent to the combined GDP of Japan, Germany, UK and France (or India) in the year 2017.

Actually, the potential cumulative addition in *gross* terms during the period 2019–2030 would be much higher-around US$ 26 trillion (i.e., equivalent to the combined GDP of the top six countries in the world excluding the US). However, making allowance for "the costs associated with managing labor-market transitions, especially for workers being left behind by AI technologies", the net incremental addition by the year 2030 has been placed at US$ 13 trillion.[4]

Notably, of the entire range of the New Age Technology underlying the Industrial Revolution 4.0 the positive impact on labor productivity of AI alone would significantly exceed the total contribution under earlier Industrial Revolutions. As noted by the McKinsey Report, "… the introduction of steam engines during the 1800s boosted labor productivity by an estimated 0.3 per cent a year, the impact from robots during the 1990s (was) around 0.4 per cent (per year), and (from) the spread of IT during the 2000s, (it was) 0.6 per cent (per year)".[5] This is in contrast to the productivity growth of 1.2 per cent per year under the ongoing Fourth Industrial Revolution.

The Report also clarifies that out of the gross incremental addition to global GDP, 41 per cent would come from the automation of labor whereas 27 per cent would emanate from innovation in products and services.

Not surprisingly, the report predicts that adoption of the emerging technology would be quite uneven both within and across countries. Within a given country, there would be 'front-runners' (around 10 per cent of companies), 'followers' (around 20–30 per cent) and 'laggards' (the remaining 60–70 per cent companies). As a result, the economic impact may be expected to become visible, gradually and become evident only over time.[6]

These studies confirm that adoption of AI in the marketplace within a country would follow a typical S-curve, a slow start in the beginning, followed by a steep acceleration as the technology matures and firms learn how best to deploy it, then tapering off in the last stage.[7]

The adoption of AI across the countries would be asymmetric as well. According to the Report,

> Best positioned, of course are China and the US, the two countries currently responsible for the vast majority of AI-related activities. Developed economies like Sweden, Singapore and Finland are well positioned to capture the benefits of AI, as well as highly motivated to do so due to their slow productivity growth. Economies with moderate foundations – think India, Italy, and Malaysia – may lag the leaders, but they have strengths in specific areas around which

they may be able to build their AI capabilities. But developing economies, which have relatively underdeveloped foundations in investment capacity, digital infrastructure and talent, risk falling further behind."[8]

In terms of specific sectors of an economy, automation and AI can play a crucial role in agriculture. In agriculture sector, many unprecedented technological breakthroughs may become possible in respect of crop health monitoring, yield production, and mapping land-use pattern. Moreover, measuring snow-melts for assessing water availability could also bring about a major transformation.

A recent study by the International Labor Organization (ILO),[9] demonstrates how "smart farming" increases productivity by using the Internet of Things (IoT), with sensors to collect real-time data which could be utilized for creating appropriate conditions that can sow, water, fertilize and harvest. Unmanned Drones used in agriculture, along with self-driving farm equipment as well as Robot pickers for fruits and vegetables are all expected to further improve productivity while reducing the need for human work.

At the same time, the emerging New-Age Technology offers greater access to product innovations, cutting-edge agricultural practices, and market developments. While commercial agriculture is likely to gain most from such innovations, smaller farmers of traditional foodgrains as well as subsistence farmers may also benefit. Notably, innovations in crop genetics, organic agriculture as well as irrigation and other infrastructure can bring about productivity improvements among small farmers.

Phone-based renting Apps for agricultural machinery such as Hello Tractor in Nigeria are enabling small farmers to access modern technology at a low cost. Small farmers could also use Apps to access agricultural extension services, as also to improve planting and crop rotation.[10] Research capacity complemented by extension and commercialization, which still remain a formidable challenge in many developing countries could also be addressed by the emerging revolutionary technology.

As to the manufacturing and services sector, the Internet has greatly enhanced connectivity, including to global markets. This has improved the development prospects for those countries that are able to take advantage of such enhanced connectivity and supply their services from anywhere in the world. In contrast to earlier strategies for economic development, which relied heavily on widening manufacturing base, services sector growth today plays an increasing role in the development process.

Emerging economies may well be able to mitigate the adverse effects of the potential setback to their exports of goods by increasing their trade in services. There could also be new and unprecedented opportunities for developing countries to engage in "remote repairing". The increasing interconnectedness of physical devices and appliances through the Internet of Things (IoT) allows technicians placed anywhere to support clients with direct troubleshooting *via* the Internet. Moreover, development of the platform economy allows developing countries to gain a larger share of global value added in services.[11]

Digitalization may also have a positive effect on the innovation capabilities of developing countries. Innovative ecosystems, such as tech hubs, placed mostly in developed countries, are accessible online and they can facilitate the exchange of knowledge and ideas among peers. Smartphone applications enable budding entrepreneurs to try out their ideas at affordable costs and risks. e-Commerce platforms allow digital start-ups to market their products to a large number of customers worldwide. According to an estimate, consumers worldwide purchased goods and services on the web, valued at US$ 2.86 trillion in 2018.[12]

In order to make the potential benefits from technological advancements more inclusive across countries, it is imperative to close the digital divide among the nation-states. A recent study[13] shows that most adults in advanced economies use the Internet whereas, several developing countries surveyed do not use the Internet. Serious gender gap in the Internet usage also persists in several developing and emerging market economies. This would have to change in the interest of harnessing the fuller gains from the emerging technology.

II. Technology-led Human Development

The New-Age Technology, especially Artificial Intelligence (AI) could potentially be used extensively for human development.

In education sector, if Public Policy ensures widespread access to the emerging technology, a 'brick and click' approach could enable considerable improvement in all four Es (i.e. Expansion, Excellence, Equity and Enterprise) of education. In many developing countries, this would help a great deal in harnessing their fortuitously available Demographic Dividend.

Consider another example, i.e. healthcare. Developing countries like India have acute shortage of medical professionals, especially physicians and specialist doctors. Looking at the rate at which new physicians or doctors are entering the system, the situation is unlikely to materially change in the near future. As a result, a large number of healthcare centers especially in rural areas have to depend only on nurses. Here, AI based on Big-Data can help in bridging the skills gap and empowering nurses.

Here's an example. While attending to a patient, the nurse enters the key symptoms and measurements into a hand-held device. Based on this information about the patient, the device provides the likely diagnosis and prescribes suitable medication. The nurse then administers a prescription that an experienced doctor could possibly have recommended, subject of course, to the consent of the patient. These recommendations are based on Big Data-AI "algorithms running at the backend—which analyze large amounts of patient records anonymously with similar symptoms and administered medications".[14] In fact, in healthcare sectors around the world, it is believed that telemedicine and bio-implants could be potentially instrumental in eradicating crippling diseases.

A discussion paper from McKinsey Global Institute[15] comprehensively deals with this issue. According to the discussion paper, AI has the potential to help address some of the world's most challenging social problems. "The existing capabilities of AI could contribute to tackling issues relating to all 17 of the United Nations (UN's) Sustainable-Development Goals (SDGs), potentially helping hundreds of millions of people in both advanced and emerging countries".[16]

The feasibility of AI deployment is truly widespread. The possibilities "range from diagnosing cancer to helping blind people navigate their surroundings, identifying victims of online sexual exploitation, and aiding disaster-relief efforts."[17] Needless to say, AI is only a part of much broader set of measures that can be used to tackle societal concerns. Combined with the other constituents of the emerging New-Age Technology, the potential of application for social good is truly enormous.

The McKinsey discussion paper[18] presents a range of applications of the emerging technology for social good, summarized as follows:

i.) **Health and Hunger:** The emerging technology can address health and hunger challenges, including early-stage diagnosis and optimization of food distribution. For example, researchers at the University of Heidelberg and Stanford University have devised a disease-detection AI system—using natural images, such as images of skin lesions to determine if they are cancerous. This diagnostic system seemingly outperformed professional dermatologists.

AI-enabled wearable devices can reportedly detect future diabetes patients on the basis of early signs, with 85 per cent accuracy. These devices could help more than 400 million people around the world afflicted by this dangerous disease.

ii.) **Educational Challenges:** In addition to significantly widening the reach of education as discussed before, the emerging technology could be immensely helpful in maximizing student achievement and in improving teachers' productivity.

iii.) **Environmental Challenges**: Sustaining biodiversity and mitigating the depletion of natural resources, pollution and climate change are important challenges that the emerging technology can address. For example, 'Rainforest Connection',[19] uses AI tools such as Google's TensorFlow in their conservancy efforts across the world.

iv.) **Crisis Response:** Distinct crisis-related challenges that include responding to human made or natural disasters, search and rescue missions as well as disease control, all of this could be addressed by the emerging technology. For example, AI on satellite data could be used to map and predict the progression of wildfires, thereby optimizing the response of firefighters. AI and data from IoT devices can also be used to create solutions that help firefighters determine safe paths through towering infernos. Moreover, drones with AI capabilities could be deployed to find missing persons in wilderness areas.

v.) **Infrastructure Management:** This emerging technology could also be deployed to address the infrastructure challenges, thereby promoting the public good in respect of energy, water and waste management, transportation, real estate, and urban planning. For example, "traffic-light networks can be optimized using real-time traffic camera data and Internet of Things (IoT) sensors to maximize vehicle throughput."[20] AI can also be utilized for predictive maintenance of public transportation systems, such as trains and public infrastructure (including bridges), in terms of identifying potentially malfunctioning components.

Having discussed, in some detail, the role of the emerging technology in promoting productivity and economic growth as well as in harnessing human development, attention may now be turned to a possible downside of the New-Age Technology.

III. Technology-led Joblessness

With the emerging New-Age Technology underlying the Fourth Industrial Revolution, the concerns regarding technology-led job displacement or joblessness have come to the centre stage.

The issue of machines displacing or replacing human labor has been discussed and debated for a very long time. However, as a problem of major interest to economists, the issue dates back to the early 1800s when David Ricardo discussed the issue in the context of the First Industrial Revolution.[21] In 1930s, the issue re-surfaced. This time it was taken up by John Maynard Keynes, who coined a new expression i.e. 'technological unemployment'. He defined it as the "unemployment due to our discovery of means of economizing the use of labor outrunning the pace at which we can find new uses of labor."[22] Keynes argued that "the increase in technical efficiency has been taking place faster than we can deal with the problem of labor absorption." He clarified, however, that "this is only a temporary phase of maladjustment".

Over the years, the debate on "technological unemployment" has been divided in two schools of thought. Optimists agree that technology may be disruptive to jobs in the short run, but they insist that there is no permanent negative impact on jobs.

Illustratively, Mashelkar, who calls himself a "dangerous optimist" stated in an interview, "In 250 years since the Industrial Revolution began, there have been continued, and sometimes very large job losses, by automation. But more jobs continued to be created and unemployment has continued to remain in single digits."[23]

At the other extreme, pessimists contend that under certain circumstances, new technologies can possibly lead to a lasting decline in the number of workers employed. Arguing that technological unemployment is more relevant today than ever before, Estela Souto says, "Whereas in twentieth century Detroit, low skilled workers temporarily lost their jobs, but were reintegrated into labor through education; the level of skills today does not guarantee employment anymore."[24] According to her, even the so-called high skilled jobs of today are likely to be victims of the technological advancements since AI and Machine Learning are soon going to have the capability to take over complex tasks. It is only a matter of time.

Perhaps, the first systematic study of this issue was undertaken by Erik Brynjolfsson and Andrew McAfee in 2011.[25] In this study, they put forth the hypothesis of "Decoupling Productivity and Employment" as illustrated in case of the US, in Chart 5.1.

Chart 5.1: Decoupling Productivity and Employment

(a)

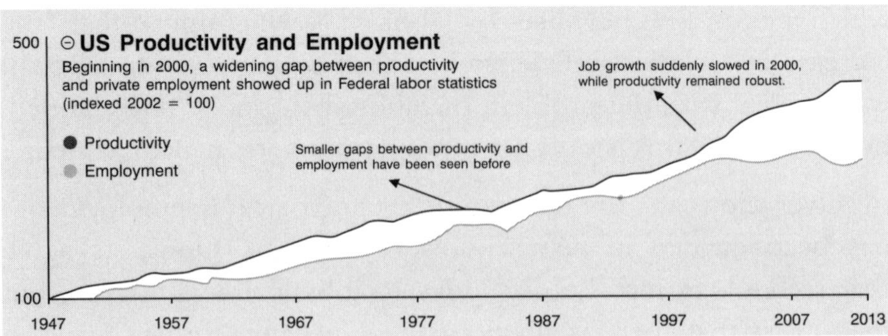

(b)

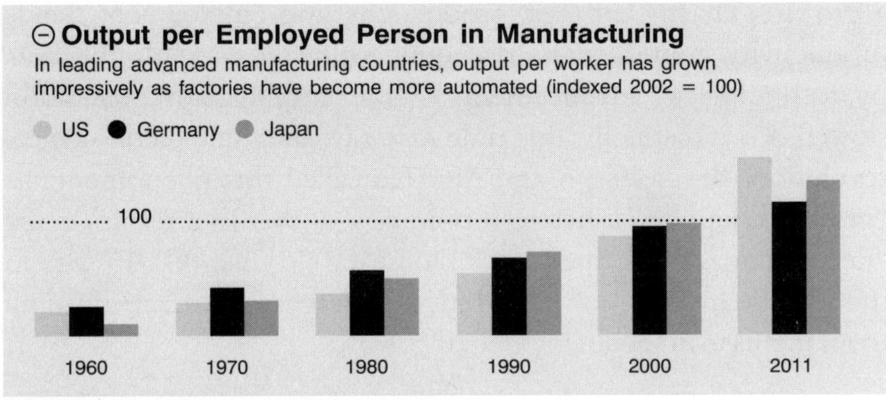

(c)

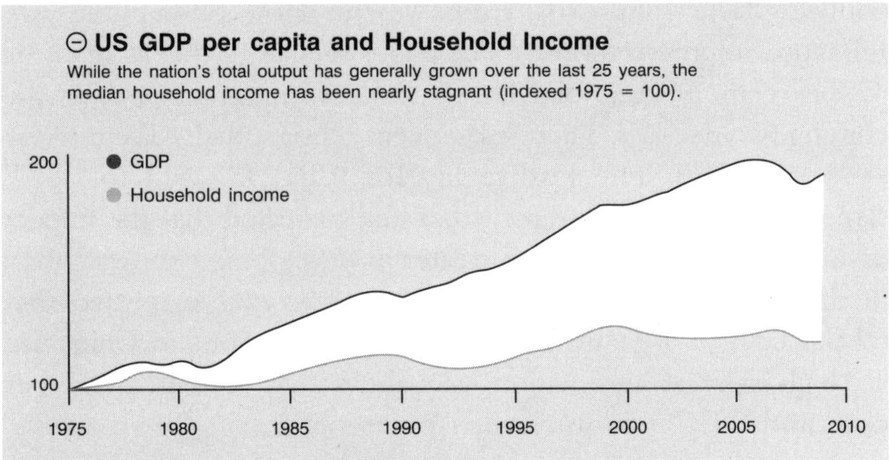

Source: Erik Brynjolfsson and Andrew McAfee

According to the hypothesis, there is co-variation between productivity and employment in the US from 1947 to 2013. As can be seen in Chart 5.1 (a), from the late 1940s to the turn of the century, productivity and employment closely tracked each other i.e. the rise in productivity was invariably accompanied by expansion in the level of employment. The underlying logic simply was that as output per employed person in manufacturing rose (Chart 5.1 (b)), the country became richer and this fueled more economic activity and created even more jobs. Then came a structural break in the year 2000. From then onwards, there was a growing chasm between productivity and employment; while productivity continued to rise quite robustly – no doubt *aided* by technological advancements – the employment generation slowed down distinctly, this time *caused* by the same technological revolution. Brynjolfsson and McAfee called this phenomenon a "great decoupling." They also pointed out that as a result of the "decoupling". While the Gross Domestic Product (GDP) rose in the US, the median household income stagnated, thereby leading to an increase in inequality (Chart 5.1(c)).

In 2013, Oxford Martin School (a community of over 200 academics at the University of Oxford) published a paper entitled, "The Future of Employment: How Susceptible Are Jobs to Computerization".[26] The paper suggested that as many as 47 per cent of US jobs are at risk of automation over the coming two decades. Their subsequent report titled, "Technology at Work v. 2.0: The Future Is Not What It Used to Be"[27] (January 2016) went a step further and indicated that the impact of automation would be far greater in developing countries than in the developed nations. The report, *inter alia,* suggested that 69 per cent of jobs in India and 77 per cent jobs in China are at "high risk" of automation, way above that of the US (47 per cent) and OECD countries (57 per cent).

A more recent report by Oxford Martin School admits that their "estimates have often been taken to imply an employment apocalypse. Yet that is not what we intended or suggested".[28] It was clarified that the research effort was intended to show that the potential scope of automation is vast, just as it was on the eve of the Second Industrial Revolution, before rendering many of the jobs that existed in 1900 redundant.

In the interregnum, two important studies, one by researchers at the University of Mannheim and another by the OECD, suggested much smaller risks for jobs exposed to automation.

Oxford Martin School has, in turn, re-examined these two reports and citing reasons, concluded that "seen through the lens of the twentieth century our estimate that 47 per cent jobs (in the US) are exposed to future automation, does not seem extraordinarily high. On the contrary, the University of Mannheim and the OECD estimates seem extremely low".[29]

On balance, there is no denying that the emerging technology is going to have a major bearing on job creation in future. Mashelkar in his interview summarizes the underlying factors at work succinctly when he states:

> To my mind, there are three (factors). First, some technologies automate physical tasks that human beings used to carry out – from typists to bank tellers, to switch board operators, to typographers. Second, some other technologies can do the intellectual tasks. These include cognitive computing, automation of knowledge work, artificial intelligence, etc. Third, the customer service tasks are also vanishing as new technologies make it possible to create self-help kiosks, or even grocery store scanners.[30]

Mashelkar indicates, elsewhere in the Interview, that autonomous vehicles will destroy the jobs of millions of drivers. Moreover, the driverless cars can have a domino effect on a range of jobs in ancillary industries including the auto insurance market, auto finance market and the like. With a deficiency in demand for these services, more jobs would invariably be lost.

The Oxford Martin School Report (2013) also pointed out that there exist large variations across countries. For example, jobs in Slovakia are twice as vulnerable as those in Norway. Broadly speaking, workers in more developed nations are seemingly at a lesser risk of losing their jobs to automation than those in the middle income and developing countries. At the same time, a wide gap also prevails between "countries of similar wealth", possibly on account of differences in organizational structure.[31]

Today, the global economy is at a critical juncture as jobs are being replaced by machines which are capable of taking decisions autonomously. For example, "we are witnessing that simple tasks of delivery, transportation and retail are being replaced by drones, self-driving cars and retails stores with no cashiers and assists".[32] With the growing automation there is a noticeable increase in layoffs across sectors, as businesses begin to realize the economic sense in adopting automation technologies.

In a *Financial Times* story, Clive Cookson points out that "Rapid strides in the development of Artificial Intelligence and Robotics threaten the prospect of mass unemployment, affecting everyone from drivers to sex workers".[33] Evidently, in this era of automation, machines have already automated most of the routine jobs in the manufacturing sector. And now, AI is learning to automate non-routine jobs as well.

It is abundantly clear that a number of AI-based technologies are coming together transforming our lives. Such technologies are enabling a synergy in multiple tasks, which are being replaced. For example, in a restaurant, one can see automation of tasks related to valet parking, doorman, checker, waiter, manager, cashier and possibly cooks. What is really happening is a thorough upgradation of the operations facilities leading to an increase in the efficiency and productivity. On the other hand, it has been engendering job losses and underscoring the need for re-skilling professionals who are at the receiving end. The only silver lining in all this is, "when physical jobs are lost due to automation, jobs for the mind open up."[34]

Since the human touch and empathy cannot be completely automated by machines, jobs would remain relevant in areas related to human interaction. People, however, would have to be skilled or re-skilled in order to appropriately adopt new technologies. In sum, skill development and skill augmentation hold the key. Whether Big Data-based AI would be a conduit for job losses or for citizen empowerment would depend on it.

IV. Technology–led New Jobs

Even while destroying a range of jobs, the emerging technological revolution will also create new jobs. There are at least three channels that one can think of how this can happen:

- Improved productivity-led new jobs
- New jobs for the New Technology
- New jobs through creative destruction of conventional jobs.

The first channel may work as follows: Wider and greater adoption of the emerging technology leads to higher productivity (as discussed in Section I). Higher productivity means lower costs, which in turn, give rise to greater aggregate demand. Larger aggregate demand provides a boost to the overall production, especially of consumer goods, thereby leading to larger employment opportunities. This channel, as explained in Section III, worked very well until the year 2000. As pointed out in the Brynjolfsson and McAfee (2011) study, there has been a 'great decoupling' between productivity and employment since the year 2000. As such, this channel may be expected to be somewhat weak.

The second channel is self-explanatory. Maintenance of the constantly evolving technology–AI and Machine Learning, Big Data and Data Analytics, Robotics, Blockchain Technology, Automated Technology, Additive Manufacturing, Internet of Things and so on, require more and more people. Moreover, people are needed to assist other people in using the emerging technology. Besides, the new technology faces obsolescence, faster than ever before. Accordingly, more people are needed to create newer technologies.

The third channel involves generation of new employment opportunities by creative destruction of the conventional jobs. Often repeated, best example here is Uber.[35] While autonomous cars are going to destroy the jobs of drivers, Uber has been generating jobs for more drivers. Uber has created a new market through a convenient and easy-to-operate reservation and payment service. Uber has, thus, enlarged the demand for drivers who, with the aid of a smartphone and a smart app, can now serve a much larger number of passengers than they would have under the traditional taxi service.

While automation has displaced a number of jobs, it has also created an entirely new ecosystem that is contributing to the creation of new jobs. Some of the most redundant skill sets, which could easily be replaced are all done away with.

Typically, jobs are composed of both "readily automatable" and "not-readily-automatable" tasks. A crucial question is: will the automation of work processes result in a reduction of the workforce, or will the "not-readily-automatable" tasks be shared among the existing workforce. The answer to this question actually depends on how work is organized in a given workplace and on the extent to which tasks that are "not-readily-automatable" can be bundled together to create additional new jobs. The automation of work processes need not present an "all or nothing" kind of mutually exclusive choice; scenario; different intermediate options do exist.[36]

As noted by Pethokaukis (2016), "one thing automation alarmists sometimes miss is that the simplistic 'machines steal jobs' story tells an incomplete tale".[37] The oft-repeated example in this regard refers to the impact of introduction of ATMs on the bank teller jobs in the US.

With the arrival of ATMs, rather than eliminating the bank teller jobs, their number actually rose, albeit, modestly. What lies behind this process of displacement and re-creation is the fall in operating and transaction costs emanating from the technological revolution that indirectly stimulates demand for labor. In the course of time, a range of new tasks are added. For instance, from a bank teller to a financial services advisor.

Pethokaukis adds: "This is in fact a much more general pattern. We see a whole number of occupations where you might think that technology is going to destroy jobs because it's taking over tasks; and the reverse happens…" For example, with improved technology in cash registers, the number of cashiers actually increased. In the 1990s when the legal profession in the US started using electronic discovery software (for searching documents) in lawsuits, the demand for paralegals increased rather than decreased.[38] It is noteworthy that in the United States, reportedly, 30 per cent of the jobs created since the late 1990s were of the kind that did not exist before, such as IT administration, hardware manufacturing, and development of smartphone applications.

The world has gone through major Industrial Revolutions before and has adopted to the new technologies, time and again. The nature of jobs has evolved, and the world has continuously witnessed the rise and fall of various kinds of jobs that were unknown earlier than become extinct.

6

Technology and Socio-economic Inequalities

"If machines produce everything we need, the outcome will depend upon how things are distributed. Everyone can enjoy a life of luxurious leisure if the machine-produced wealth is shared, or most people can end up miserably poor if the machine-owners successfully lobby against wealth re-distribution. So far, the trend seems to be towards the second option, with technology driving ever-increasing inequality."

Stephen Hawking

Introduction

While the emerging technological revolution that is building up the Industrial Revolution 4.0 has been transforming human lives positively as never before, it has also been posing several challenges. This chapter is focused on the possible downsides of the emerging technology, especially, in terms of its potential adverse impact on socio-economic inequalities. The chapter also explores the likely effect of widening inequalities on democratic governance around the world.

The chapter has been organized in three sections as follows:

Section I provides a primer on socio-economic inequalities, especially income (and wealth) inequalities, in terms of the basic concepts, measures, received theories and the state of affairs in this regard around world *prior* to the emergence of the New-Age Technology. Against that backdrop, Section II discusses the possible nexus between the emerging revolutionary technology and socio-economic inequalities. Finally, Section III dilates on social implications of upheavals in the job ecosystem.

I. Socio-economic Inequality: Concepts, Measures, Theories and Status

A. Socio-economic Inequality: Concepts

Socio-economic inequality essentially refers to differentiation made in a society in terms of availability of social goods and economic resources, which have a bearing on the well-being of people. Obviously, socio-economic inequality is a combination of social and economic inequality.

Of the two inequalities, economic inequality amounts to an unequal distribution of income, wealth or standard of living. Social inequality on the other hand, is not as straight-forward to define. It covers differentiation made on the basis of a range of factors such as caste, race, religion, ethnicity, gender, sexual preferences and social class. It is this differentiation that takes the form of uneven distribution of social goods such as education, healthcare and entitlements including freedom of speech, political representation and participation and overall social inclusion.

B. Measures of Inequality

Here again, while measuring economic inequalities is quite straight-forward, assessing social inequality is not.

There are four commonly used measures for economic inequality: (i) Average Income (ii) Percentile Ratios (iii) the Gini Co-efficient and (iv) the Theil Index.

Average Income as a measure of inequality is self-explanatory. Typically, median income is the simplest measure used. Percentile Ratio measure of income inequality looks at a given income distribution and computes the ratio of income of households in the 90th percentile to the household in the 10th percentile. The Gini Co-efficient or Gini Index is yet another measure whose value varies generally, between zero and one. A Gini Co-efficient of zero implies perfect equality (i.e. no inequality in the income distribution) whereas the value of the index at one signifies the most unequal distribution of income. In other words, value of the Gini Index is directly proportional to the extent of inequality. Finally, the Theil Index as a measure of income inequality is somewhat involved. A distinguishing feature of this index is that it allows decomposing changes in income inequality into within-country and between-country changes, which is evidently useful in cross-country comparisons.

C. Theoretical Models of Inequality

In a seminal and pioneering work, way back in 1955, Professor Simon Kuznets put forth a hypothesis that income inequality initially rises (as an economy makes a transition from the agrarian state to an industrialized one), reaches its peak and then declines as the services sector of the economy comes to dominate. The initial rise in income inequality emanates from higher average household incomes. Over time, however, as labor market developments (and often, compensating public policy actions) exert their influence – the so-called 'mitigating effects', the income inequality comes down.

This phenomenon was illustrated using an inverted u-shaped curve (with per capita income on x-axis and income inequality on y-axis) which came to be called the Kuznet's curve.[1]

While Prof Kuznets received the Nobel Prize in Economic Science in 1971, another Nobel laureate in the same discipline (in 1969), Jan Tinbergen arrived at a similar result in 1970.[2] Prof. Tinbergen analyzed the effects of technology on labor markets. According to him, technological change enables highly skilled workers to get a premium, leading to increased income inequality. Overtime, when the labor market undergoes adjustments, the premium tapers off and hence, the income inequality comes down.

Thomas Piketty, a French Economist came out with a book in 2014 on the subject that was hailed as "a political and theoretical bulldozer".[3] Another Nobel laureate, Prof Robert Solow, summed up the book succinctly as: "a new and powerful contribution to an old topic: as long as the rate of return (on capital) exceeds the rate of (economic) growth, the income and wealth of the rich will grow faster than the typical income from labor".[4]

According to Piketty, when the rate of return on capital (r) (which includes profits, dividends, interest, rents and other income) exceeds the rate of economic growth (g), wealth tends to accumulate quickly from capital rather than labor. When wealth accumulates faster among the top 10 per cent and top one per cent of wealth-holders, income inequality increases sharply.

Piketty argued that inequality is an essential feature of capitalism, which can be contained only through the State intervention, such as a progressive wealth tax (of say 2 per cent) combined with a progressive income tax (as high as 80 per cent), if the government is serious about reducing inequality.

Piketty's long period data showed that the average return on investment outpaces productivity-based income by a wide margin. Accordingly, he argued that bursts of productivity emanating from technological advances (no matter how transformative they are), cannot be depended upon to ensure sustained economic growth path. According to Piketty, in the absence of major tax reforms (that he proposed) the global economy would be characterized by a combination of low economic growth and high inequality. Expectation of a just and rational order by itself (i.e. without governmental intervention), would be futile.[5]

None of these theoretical models can offer a reasonable explanation of the pattern of income distribution over longer time periods. As a compromise, Milanovic has proposed a Theory of Kuznet Waves according to which, "technology, globalization and public policy co-evolve to generate successive cycles of increasing and decreasing income inequality. These cycles are not fully synchronized across countries and regions and their duration varies".[6]

D. Status of Inequality: Global and National

Before discussing the potential impact of the emerging New-Age Technology on global and national inequality, it is important to have a perspective of the ground realities i.e. the actual long term variations in income inequalities.

The picture emerging from some recent studies depicts the following:

i.) Global income inequality increased almost unabatedly from the First Industrial Revolution upto the 1980s. For example, according to Bourguignon and Morrison study[7] (2002), in a sample of 33 countries, the Gini Co-efficient rose steadily from around 0.5 in 1820 to about 0.66 in the 1980s.

ii.) This upward trend in global inequality was punctured, albeit briefly, only after World War II.

iii.) In the 1980s came the turning point when income inequality started coming down.

Another study covering a much larger sample of nation-states came up with similar results. According to Milanovic (2012) study "after a historical increase, inequality had been relatively stable beginning in the 1960s and then started to decline in the early 1990s,"[8] from a Gini Co-efficient of 0.63 to 0.53 in 2010. According to the study, "this reversal is strongly influenced by rapid growth and increased household incomes in emerging (market) economies, not only China and India, but also Indonesia, Bangladesh, Kenya and Nigeria".[9]

Inequality within nation-states surprisingly presents a much diverse picture as per the World Bank and OECD data.[10]

Conclusions emerging from the World Bank and OECD based data of 122 countries, may be summarized as follows:

Depending upon the underlying pattern, the 122 countries could be divided into four district categories, as follows:

a) Inequality increased persistently

There were 52 countries in this category including Australia, Canada, China, Spain, France, Indonesia, India, Ireland, Israel, Italy, Japan, Mexico, New Zealand, Norway, Sweden, UK and USA.

This category comprised several advanced economies and fast-growing emerging market economies.

(For example, the Gini Co-efficient in the US rose from 0.369 in 1986 to 0.411 in 2012, whereas in China, it rose from 0.299 in 1987 to 0.422 in 2012).

Looking at it somewhat differently, among the OECD group of countries, between 1985 and 2012, real disposable income of the bottom 10 per cent of households rose by 14 per cent whereas the same increased by 50 per cent for the top 10 per cent.[11] Wealth inequality among the OECD countries was much sharper. The top 10 per cent households held 23 per cent of the income but more than 50 per cent of private wealth. [12]

b) Inequality increased initially, followed by decrease

There were 35 countries in this category including Argentina, Austria, Brazil, Thailand, Tanzania, Uganda, Ukraine and Uruguay.

(For example, the Gini Coefficient for Brazil rose to 0.601 by 1993 and after remaining at that elevated level until the turn of the century, it declined to 0.527 in 2012).

c) Inequality decreased early on, followed by increase

There are only 8 countries in this category which, *inter alia*, include, Russia, Pakistan, Venezuela and Zambia.

d) Inequality decreased consistently

There are 27 countries in this interesting category, which include, Switzerland, Germany, Iran, Cambodia, Sri Lanka and Nepal.

II. Nexus Between Technology and Socio-economic Inequalities

As elaborated in Chapter 5, the emerging technology that underlies the Industrial Revolution 4.0 is throwing up immense opportunities and possibilities. The revolutionary technology could potentially give a significant boost to labor productivity and in the process, could generate enormous additional wealth for many developed as well as emerging market economies.

On the other hand, as pointed out in Chapter 1, while the global economy could be on a high growth trajectory, it is likely to come with a major cost. One of the most important downsides, it is argued, could be a sharp increase in socio-economic inequalities. This in turn, could weaken the social fabric or could even threaten the social stability within the nation-state concerned. It is generally agreed that high and rising levels of inequality are likely to jeopardize or at least weaken the cohesion and working of democratic societies.

What is the nexus between the emerging revolutionary technology and the socio-economic inequalities within and across countries?

Arguably, the relationship is a circular one. Illustratively, the evolving technology may aggravate socio-economic inequalities. The growing inequalities in turn, may exacerbate the so-called "Digital Divide",[13] which in its own turn, may bring about a further deterioration of socio-economic inequalities. Importantly, the circle does not necessarily have to be a vicious one. With appropriate Public Policy interventions, it could as well be a virtuous circle – running from technology to reduced inequalities to less severe "Digital Divide" and on to a further reduction in inequalities.

A. Causation from Technology to Inequality

This causal relationship has been explored by several studies. Most studies, however, seem to conclude that the impact of the emerging technology on socio-economic inequalities is ambiguous.[14]

As discussed in Chapter 5, the emerging revolutionary technology has a very significant bearing on the jobs. There is fervent hope but only limited corroborating evidence that the New-Age Technology could generate a sufficiently large number of new jobs so as to compensate for the higher productivity-induced job losses.

The received literature on the subject maintains that the emerging revolutionary technology could affect income inequality in "multiple and often contradictory ways".

First, with the New-Age Technology there would be a big demand for high-skill jobs. Acquisition of the high-end technical skills under the emerging technology may be affordable only for the rich. Even if more low-skill, low-paying jobs are created simultaneously, this would drive a larger wedge between the haves and have-nots thereby, creating greater socio-economic inequalities.

Secondly, at the other end of the spectrum, the New-Age Technology could potentially be more inclusive. It could enable those individuals who were excluded from the marketplace earlier, to participate in the production process. According to Krugman (1979), high-income countries develop new technology and initially produce goods and services using such advanced technology. Once low and middle-income countries catch up with the technology, they deploy it to produce and export those goods and services. Since labor is typically cheaper in low and middle income countries, they remain competitive, even though wages in these countries increase over time.

Needless to say, the working of this phenomenon depends entirely on the ease and speed with which the new technology is replicated and used by low and middle income countries. The emerging technology tends to accelerate technology transfer because it (i.e. technology) makes it easier to outsource and off shore high-skill production. As capital moves across borders faster and new knowledge is shared more rapidly, the ability of high-income countries to protect higher wages is eroded while the low-income countries benefit from wage increases.

Moreover, given the characteristics of the emerging technology such as, modular design, the end-to-end principle of connectivity embedded in its architecture, high rates of innovation and the plasticity of digital technology, a tremendous wave of innovations, new industries and employment opportunities could be unleashed. This could tend to reduce the socio-economic inequalities, particularly if it increases incomes below the median.

The impact of the emerging technology on inequality is ambiguous essentially because it is not clear, *a priori*, whether the inequality-reducing forces would outweigh the inequality-increasing factors. At the same time, there is a broad agreement that the set of factors which would dominate depends upon the country-specific economic, technological and institutional context in which they unfold, as well as the stance of the Public Policy toward socio-economic inequalities.

In this regard, research studies[15] draw three main scenarios:

- Inequality Reducing
- Inequality Enhancing and
- Maintaining Tolerable Inequality.

a) Inequality-reducing Scenario

This scenario is said to be most likely under the following conditions:

i.) The education and skill development levels of the labor-force are high;

ii.) The education and skills ecosystem is continually reviewed, revised and updated;

iii.) The adaptability of labor and enterprises to the changing market conditions is high, and

iv.) Appropriate public policies are in place with a view to mitigate the socio-economic inequalities.

b) Inequality-enhancing Scenario

Under this scenario, exactly the opposite is applicable. The state of education and skill development is unsatisfactory, serious efforts for improvement in the education-skill ecosystem are lacking, adaptability of workers and businesses to rapidly changing business environment is underdeveloped and there is inadequate Public Policy support to contain or mitigate socio-economic inequalities.

c) Tolerable Inequality Scenario

Under this scenario, with technology advances socio-economic inequality within a country does increase, but it is accompanied by a steady improvement in average or median per capita income or enhanced labor market participation.

What level of inequality is tolerable is a moot point. Under ideal circumstances, no increase in inequality should be "tolerable" given its harmful effect on democracy. On the other hand, it has also been argued that what really matters is *not* the extent of inequality, but whether the lowest income strata of the society is able to support a decent standard of living.[16] That is precisely where "universal basic income" concept is being construed as an important Public Policy initiative, especially in low and middle-income countries.

B. Causation from Inequality to Technology

The "Digital Divide" could be conceptualized by the subject that needs connectivity – i.e. individuals, organizations, businesses and countries. It could also be seen in terms of the attributes of those subjects, such as, income, education, sex, geographic location (rural-urban), etc. Across the world, countries could be characterized by different dominant attributes.

To illustrate, in an emerging market economy like India dominant attributes of the "Digital Divide" could be two: first, the rural-urban divide – urban households have significantly larger access to the Internet than those in rural areas. Secondly, the gender-divide. Reportedly, "while half of all Indian men have a mobile phone, fewer than one in three women have one".[17] While the literacy rate was 80 per cent among men, it was less than 66 per cent for women as per the Census of India (2011). In other words, rural population, women as well as the marginalized strata of the society are placed at the wrong side of the Digital Divide in India.

In this digital age, there is no doubt whatsoever that an individual should be able to connect in order not only to benefit from the formidable economic gains in productivity, but also for effective social and cultural participation for self-fulfilment. Understandably, access is a necessary condition for overcoming the Digital Divide, but by no means is it sufficient. In other words, by itself, access to connectivity cannot enable the individual to realize his/her socio-economic and cultural goals. What is needed, in addition, is adequate digital infrastructure and ability to comprehend the information provided by the Internet.

In other words, prevailing inequalities among various socio-economic groups in a country, can create barriers to access and effective deployment of the emerging technology. Clearly therefore the chain of causation not only could run from technology to inequality, but *vice versa* as well i.e. from inequality to technology also, thus completing either the vicious or virtuous circle. This is where the stance of Public Policy assumes crucial importance.

III. Social Implications of Upheavals in the Job Ecosystem

With the emerging revolutionary technology, it is clear that there will be a pendulum swing in the global job ecosystem. Jobs will come and go and hopefully, job creation will outweigh jobs losses. But the focus will remain on efficiency, higher productivity and greater value addition in the overall level of human performance. Needless to say, policymakers must be prepared for the worst and should always be mindful about ensuring that everybody makes a living. The world history is replete with examples of how high unemployment can lead to social unrest, civil uprising and political upheavals that nation-states would like to prevent at any cost.

Imagine a situation in future when most of the work is automated, thanks to the emerging revolutionary technology. The automated work would include farming, personal transportation, goods transportation, maintenance of roads, generation of power, manufacturing, schooling, medical care, elderly care, customer service and so on. Not that humans will be, or even could be, completely eliminated from the equation. It is only that their role will become either minimal or move up the cognitive scale; they would function as casual supervisors or auxiliary helpers when things get really tricky, and that too on very reduced work hours. As William Gibson said, "The future is already here – it's just not very evenly distributed".[18]

Rather than outright and widespread job losses, what might happen is a gradual reduction of work hours, eventually moving employees from full-time to temporary positions. The loss of pay need not hurt because there will be a corresponding decline in costs for all types of goods. "The general direction will be towards less job-related work without any corresponding reduction in the quality of life (and perhaps an overall improvement in the quality of lifestyle). This progress will obviously not happen in a straight, happy path, but will meander agonizingly, causing considerable distress, backtracking and excitement along the way. There will be a transition, which despite all the hype and exuberance, will have a fairly long trajectory.[19]

In the course of time, hard work will be replaced by 'smart' work. Financial freedom will give way to time freedom.

The emerging technology may cause job losses, but that will be superseded by the loss of meaning in traditional work and the virtue of duty-bound toil, leading for people to re-examine their motivations and inspirations for doing what they do.

Increasingly, the emerging New-Age Technology would allow firms to save money by hiring freelancers and independent contractors instead of permanent workers. This has already created the so-called Gig Economy – a labor market characterized by short-term contracts and flexible hours at the cost of irregular jobs with lesser benefits. Reportedly, in 2016, one in three workers were employed in the Gig Economy, but not all by choice. Policymakers must, therefore, take up massive skill development and up-skilling programme.

According to Moshe Vardi, many people will be unpleasantly shocked to find they have no or limited passion and motivations for what they do in life. For them, "AI spells debilitating inertia, deepening despair and the slide into apathy. A happy group (hopefully, more than a few) will look upon this as a golden opportunity to get rid of the shackles of work they must do so they can do work they want to do."

While the global economy would be on high growth trajectory, economic disparities and social inequities are likely to rise, thereby weakening or even threatening the social fabric. One possible reason is that acquisition of the high-end technical skills under the emerging technology may be affordable only for the rich. Another plausible reason why this might happen is the likelihood of high costs of organ replacements, and artificial bio-implants. Only the rich will be able to afford such costly healthcare. Consequently, the rich may become physically stronger and mentally more agile than the poor, thereby driving a larger wedge between the haves and have nots. An extreme possibility is the super-rich, biogenetically engineering 'perfect designer babies' (this is what Edward Wilson called God-like technology), thus creating a new form of social stratification like the caste-system.

The AI revolution will be eclipsed by the revolution caused by decoupling jobs from work and by moving from "I work, therefore I live" to "I live, therefore I work". Sociologists will be forced to re-examine and re-formulate their models of human interaction and organization. Economists will be forced to invent new rhetoric for their platforms when the traditional political posturing will become moot. Schools will be forced to contend with a de-schooled society.

Society as a whole must then grapple with the deeper social, economic and psychological ramifications of permanent net job losses caused by the New-Age Technology. A possible silver lining is that the loss of jobs without loss of lifestyle should give humans the time and freedom to contemplate on these issues, in their own enlightened collective self-interest.

7

Technology: Banking-Finance 4.0

"Technology always has, and always will, spur innovation in finance. The question is whether this innovation will benefit all, or only select few. If handled correctly, fintech will cut the cost of utilizing financial tools and enable millions to fulfill their aspirations of building a better life. That is why, I believe it is our shared responsibility to create a safe, sound, sustainable and inclusive financial system, protected from criminal abuse."

Christine Lagarde[1]

Introduction

Emerging technology underlying the Industrial Revolution 4.0 has been disrupting practically every sector of economies. Indeed, the pace of disruption is steadily accelerating. Financial services sector is no exception. Arguably, banking and financial services industry is currently going through a metamorphic transformation, although the extent and pace of change differs widely across the world. This chapter is focused on the impact of revolutionary technological advances on banking and finance, leading them to what is being referred to as Banking 4.0 and Finance 4.0, in step with the Industrial Revolution 4.0.

This chapter has been organized as follows:

Section I traces the historical evolution of Banking-Finance 4.0. Section II discusses various dimensions of the emerging Banking-Finance 4.0. Against that backdrop, the final section i.e. Section III dilates on the future of financial services including e-Commerce and crypto-currencies.

I. Evolution of Banking-Finance 4.0

As indicated in Chapter 1, First industrial Revolution (1750-1870) was brought about by the use of steam powered engines and was marked by the development of iron and textile industry. It was certainly no coincidence that the early central banks – the Swedish Ricks bank and the Bank of England – were established *before* the beginning of the Industrial Revolution, in the years 1668 and 1694 respectively.[2] This was nothing but Banking-Finance 1.0 corresponding to the First Industrial Revolution.

The Second Industrial Revolution (1870-1914) was engineered by the use of electricity (instead of steam) for mass production. New industries included steel, oil and electricity generation; light bulbs and telephones were invented. This Industrial Revolution, not surprisingly, coincided with massive colonization drives, particularly into Asia and Africa. The Industrial Revolution 2.0, in turn, was greatly facilitated by the rise of credit markets and development of stock and bond markets, which may, therefore, be called Banking-Finance 2.0.

The Third Industrial Revolution (1980 onwards), as noted before, was engendered by the Internet and information and communications technology (ICT). Both, the industry as well as banking and finance sectors expanded rapidly everywhere with the invention of personal computer (PC) and use of digital technology. In step with the Industrial Revolution 3.0, there was Banking and Finance 3.0, characterized by the emergence of so-called financial derivatives. As Andrew Sheng, a noted former Bank-Finance regulator put it: "Finance 3.0 was the age of financial derivatives, in which very complex (and highly leveraged) derivatives became so opaque that investors and regulators realized they became what Warren Buffett called 'weapons of mass destruction".[3] Finance 3.0 came to an abrupt halt in 2008 when there was a global financial crisis, "global meltdown" as it was then called. Finance 3.0 was revived only after massive bail-out packages (in the US, Europe and China) and unusually large central bank interventions worldwide.

As explained in Chapter 1, Industrial Revolution 4.0 currently underway, involves several exponential technologies such as Artificial Intelligence (AI), Machine Learning, Big Data Analytics, Cloud Computing, Internet of things and the like. The confluence or fusion of these cutting-edge technologies "means that cyber-physical systems can interact with each other to produce, distribute and trade across the world in a massively distributed system of production".[4]

With one-to-one correspondence again, *pari-passu*, Bank-Finance 4.0 is emerging, as discussed next.

II. Dimensions of Emerging Banking-Finance 4.0

Broadly speaking, financial system of any nation-state traditionally comprises the banking system–at its core, and non-bank financial sector, which includes equity, bond and commodity markets (technically also called the shadow banking system), besides the insurance sector. The conventional view is rapidly changing. Increasingly, the traditional business model of banks and insurance companies is being challenged, even threatened, by the so-called "FinTechs" and "TechFins" which include a wide range of financial service providers and start-ups. Empowered by Big Data Analytics and Artificial Intelligence, the FinTechs and TechFins are able to offer consumers "easier and more intuitive access to financial services" and in the process, technology giants with "enormous customer bases and deep pockets" are "causing disruption to the global financial system".[5] It is through this metamorphic disruption that Banking-Finance 4.0 seems to be emerging.

A. Bank 4.0

The immediate challenge that banks are facing worldwide is the need for massive digitalization. Banks in many advanced and emerging market economies have increasingly been relying on process automation with a view to reduce costs and expedite delivery of financial products and services. With the emergence of FinTechs and TechFins, however, banks are feeling the heat. They are under pressure to offer more and newer digital options in order to cope with growing customer expectations.

According to Brett King, the author of the seminal book on banking, *Bank 4.0: Banking Everywhere, Never at a Bank*, "Friction will be the biggest killer of bank revenue in the next 10 years…. The lowest-friction experience will be most widely adopted, and if that is *via* a FinTech's app, consumers won't look in the rear-view mirror at the bank process they used to rely on."[6]

Regrettably, as King points out: "There are banks that are still fundamentally Bank 1.0 operationally and in customer engagement". According to him, it is appalling that, "The majority of banks still don't offer account opening on a mobile, and thus would barely qualify for Bank 3.0 status – sort of Bank 2.5".

What is imperatively needed for banks according to King is omni channel digitalization. "Customers expect a seamless experience across all channels. To be able to offer that, banks need to digitalize a whole range of processes, including back-office operations". In this regard, Robotic Process Automation (RPA) has been recommended as an effective option for automation of several processes.[7] This important step could be very useful *inter alia* in (i) speeding up the processing of Big Data, (ii) upgrading back office processes for faster execution with fewer errors and (iii) freeing up employees to focus more on clients and improve customer service.

While massive digitalization of the traditional way of doing things is necessary for banks, by itself it is not sufficient. As Brett King explains, what is really needed in addition, is "reimagining banking". Making banking processes "frictionless" (through automation etc.) is just the beginning. What follows that is making them "contextual". In the none-too-distant future, according to King, "banking will increasingly become an embedded function"[8] in the value chain of experience. "The banking part of purchases and other events ceases to have a separate existence, and, ideally, get reimagined".[9]

King insists that re-imagining banking is very different from merely digitizing the old way of doing things. According to King's vision, a Bank 4.0 institution engages in "first principles design thinking", which involves "backing up from the task, tearing up the old blueprint, and confronting the job afresh".[10]

Gone would be the branch-based delivery models. Also gone or diminished would be product-based banking. In his futuristic scenario of "embedded" banking, illustratively, credit cards would be replaced by "predictive and contextual credit access," overdrafts will become "emergency credit access" while checking account, current account or debit card will take the form of "cloud-based personal value store linked to a mobile wallet".[11]

Banks, more than other financial institutions, will be subject to 'Technological Darwinism'. Fundamentally, there is an ongoing race between banks and non-bank financial institutions – both aiming at eliminating friction in people's access to financial services. Any friction that isn't absolutely essential will have to be eliminated using the emerging technology. If banks fail in removing friction, non-bank financial institutions will do that. Arguably, as King seems to think, it is only "contextual banking" combined with the "first principles design thinking" that will help banks survive in an operationally meaningful manner.

B. Finance 4.0

It is the advancement towards Finance 4.0 that has been driving banking systems everywhere towards Bank 4.0. A pertinent question then is what is driving the emerging Finance 4.0?

The underlying reasons are not far to seek. In the aftermath of the global meltdown of 2008, which in severity was next only to the Great Depression of 1930s, "the amount of financial regulation have tripled in terms of number of rules and complexity on what the official sector (i.e. government and the central bank) can regulate, which is mostly the banking system".[12] Not surprisingly therefore, "all the innovation, talent and money are moving to outside the banking system into the asset management industry, which is much more lightly regulated".[13]

According to Sheng, "all those who wanted to avoid regulation, taxation or any official oversight moved their trading and transactions off-balance sheet and also off-shore".[14] This "shadow banking" system is essentially the non-bank financial system which includes stock, bond and commodity markets. However, Sheng points out that the bulk of illegal, illicit transactions traditionally was done in cash".

Emerging technology underlying the Industrial Revolution 4.0 has had tremendous impact on financial services sectors around the world. Seemingly, there are at least three causal links:

i.) The increasing use of Robotic Process Automation (RPA) for eliminating human errors,

ii.) The experimentations around cyber currencies or cryptocurrencies and most importantly,

iii.) Peer to peer transactions driven by a distributed ledger system – i.e. the 'Blockchain technology'.[15]

Finance 3.0 and its earlier versions were based essentially on a hierarchical or top-down ledger system wherein financial settlement between any two parties were taking place on a ledger of higher order.[16] Illustratively, payments, say from person X having an account in bank A to person Y with account in bank B, were settled in domestic currencies across the books of the respective central bank. Likewise, in international trade and payments, settlements were mostly done in US dollars (being a reserve currency) across the ledger of the US Federal Reserve Bank.

As mentioned above, the basic difference between Finance 3.0 and Finance 4.0 lies in the underlying architecture. Finance 3.0 is characterized by a hierarchical pyramid-like ledger whereas Finance 4.0 is distinguished by the underlying distributed ledger system emanating from the Blockchain technology.

Finance 4.0 is flourishing in the financial environment around the world characterized by at least three principal factors:

i.) Virtually unipolar world order is changing towards a multi-polar world. Traditionally, the US dollar was the only reserve currency in which international trade was invoiced and payments were made. It served as the international medium of exchange, store of value as well as unit of account. Over the years, this has changed dramatically with Euro, Pound-Sterling, Japanese Yen and Chinese Renminbi added to the list. Moreover, a number of cryptocurrencies like Bitcoin have come up. (Issues relating to Cryptocurrencies have been discussed at length below).

ii.) The traditional top-down architecture is now merging with a bottom-up architecture in which trade, transactions and stores of value are shifting towards the PtoP, the shadow banking system.[17]

iii.) The cyber – PtoP business is advancing because the official sectors (i.e. governments and central banks) are concerned that further regulation might restrain the stream of innovations.

As a combined result, Finance 4.0 may be expected to trigger the fastest metamorphic change in financial systems. FinTechs and TechFins are likely to "upturn centuries of tradition by challenging legacy institutions, redefining the idea of trust in the banking system, bringing crowdsourcing, collaborative and self-serving processes to the centre of a slew of new services, and delivering on highly elevated levels of customer service expectations".[18]

III. Future of Financial Services

How is Banking-Finance 4.0 likely to evolve in the future? Equivalently, what would be the distinct dimensions of the future growth and development of the financial services sector around the world?

In this regard, one can think in terms of following three aspects:

i.) Competition between Banks and non-Bank Financial Institutions,

ii.) Growth and Development of e-Commerce, and

iii.) Growth and Development of Cryptocurrencies.

A. Competition between Banks and non-Bank Financial Institutions

At the outset, the difference between banks and non-bank financial institutions such as FinTechs and TechFins needs to be clarified. Banks are traditional financial intermediaries. They intermediate, i.e. transfer financial resources in any economy from net savers to net borrowers, i.e. from those who spend less than they earn to those who earn less than they spend.[19]

At the other end, there are non-bank financial institutions, comprising FinTechs and TechFins. The difference between the two is essentially based on the origin of the underlying organization. FinTech typically "references an organization where financial services are delivered through a better experience using digital technologies to reduce costs, increase revenue and remove friction,"[20] for example, PayPal in the US. Alternatively, a TechFin "usually references a technology firm that finds a better way to deliver financial products as part of broader offering of services".[21] Examples here include Google, Amazon, Facebook and Apple in the US and Alibaba and Tencent in China.

Jack Ma, the Founder of Alibaba said in an interview at the 2015 World Economic Forum:

> "I said we will make a bet: in 10 years we will be bigger than Walmart, on sales. If you want 10,000 new customers, you have to build a new warehouse and this and that. For me?....two servers".[22]

This is symptomatic of the potential of TechFins and FinTechs. A frequently asked question about the emergence of FinTechs is "Whether they are here to replace the banks?" The answer is, they are not. FinTechs are not meant for killing the banks but they are here to change the meaning of financial services itself. And in this process, if some banks fail to put their act together, then yes, FinTechs are going to replace them. "Every Fintech in the world has the same mission. Kill not the banks, agents, brokers and insurers of the world, but *kill the friction associated with financial services today"*.[23] Banks, on the other hand, have to face their traditional constraints including compliance based resistance and legacy issues.

Not surprisingly, today the fastest growing financial services are coming not from banks but from the FinTechs, TechFins and emerging bank start-ups. According to a Ernst and Young survey conducted in 2019, adoption of FinTech services has moved steadily upward, from 16 per cent in 2015 to 33 per cent in 2017, to 64 per cent in 2019. "Worldwide, 96 per cent of consumers (among the Internet enabled population) know of at least one alternative FinTech service available to help them transfer money and make payments".[24]

According to Brett King, standards by which banks are operating are no longer their own – "it's more than likely that the day-to-day banking experiences you enjoy in 2025 will have been most heavily influenced by technology startups than by incumbent banks who have innovated. FinTechs and TechFins are shaping the landscape that is to come. Not regulators, not banks".[25]

The Ernst and Young survey mentioned above and its earlier versions, covered customers who were digitally active and found that most of them chose to use FinTech because of the ease of access. Basically, the ease of opening an account which FinTechs provide is the primary motivating factor for customers to switch towards these platforms. According to King, no customer wants to go to a bank for opening an account. The lack of effort by the banks to reduce the current friction in opening accounts is a good example of how the FinTechs are emerging as a global player preferred over banks. For banks to compete with the FinTechs, they would have "to let go the requirement to sign the application form with a wet signature".[26]

A report by Accenture in 2015, after analyzing the impact of FinTechs on the banking industry, concluded that there are only two possible options for the banks (due to the emergence of the FinTechs), "Either, they continue their product-based sales approach and continue to lose their profits to the emerging FinTechs or they integrate innovations at the business model level improving the customer experience".[27]

As per a McKinsey Report (2016), only 8 per cent of the banks surveyed believed that they need *not* respond to the emergence of the so-called FinTechs.[28] This means that majority of the banking sector knows that they have to compete with these firms at present and in the future.

Yet another report, by PricewaterhouseCoopers (PwC) observes that "Disruptors are fast moving companies, often start-ups, focused on a particular innovative technology or process in everything from mobile payments to insurance".[29] And, they attack some of the more profitable elements of the financial services value chain. The survey found that a quarter of the businesses could be at risk of losing to stand alone FinTech companies within five years.

The PwC Report also discusses the notion of "sharing economy" defined as a decentralized asset ownership, wherein IT is used to find efficient matches between providers and users of capital, rather than automatically turning to banks as an intermediary.

Brett King sums up the issue as follows: "Ultimately banks will have only four choices in responding to the technological shift. First, they do nothing and slowly decline in obsolescence. Secondly, partner with FinTechs that will be cheapest and fastest. Thirdly, acquire a FinTech, which will be fast but very expensive. And fourthly, copy FinTech innovations which will be slow and expensive".[30] The second option seems to be emerging as the most relevant one in the current scenario. Notably, in 2012, more than 50 per cent of banks felt that FinTechs were largely irrelevant but by 2017, 93 per cent of them felt that there is a need to partner with the FinTechs.[31]

B. e-Commerce

The dramatic change in Information Technology over the last decade or so has completely transformed the way people generally do their shopping. The prevailing system gives a choice to the customers to purchase goods without actually going to the shop. Instead, they carry out the desired transaction through an online platform specially designed for this purpose. In other words, e-Commerce is buying and selling of goods and services on the online platforms.

The e-Commerce system can be divided into three models:

1. Business to Business (B2B)

B2B implies selling by one business entity or manufacturer to another business or industry.

2. Business to Consumer (B2C)

This type of commerce is widely used among the online platforms. Here, any business entity directly sells its products and services to the consumer. In this model, the customer directly views the products on websites and order it online. After the order is received, the selling company processes it and delivers it to the buyer. Amazon, Alibaba, Flipkart, etc. are some of the major e-Commerce giants across the globe.

3. *Consumer to Consumer* (C2C)

In this model, consumers sell the product and services directly to other consumers through various platforms. Platforms such as OLX, Quikr, etc. are some of the examples that are based on this model.

i.) Global e-Commerce

Global e-Commerce involves selling products or services across geopolitical borders from a company's country of origin. As per the projections of Digital Commerce, worldwide purchases by consumers rose from US$ 1.3 trillion in 2014 to US$ 2.86 trillion in 2018 and are expected to shoot up to nearly US$ 5 trillion by 2021.[32]

The global market of online retailing is dominated by the world's two largest economies, i.e. US and China. These two countries accounted for more than half of total worldwide sales of physical goods in 2017. China, the leading e-Commerce market, saw online sales increase from US$ 685 billion in 2016 to $877 billion in 2017, almost 28 per cent increase in the e-Commerce business.[33]

According to Bain & Company's Annual Luxury Study, Asia is the "growth engine" for online luxury sales.[34] The category jumped an impressive 23.3 per cent in 2017. Here again, consumers in Mainland China are leading the way.

China has outperformed the United States in e-Commerce in terms of sales as well as growth. In the first quarter of 2019, total retail sales of US were estimated at US$ 1,344 billion, which was virtually unchanged from the fourth quarter of 2018.[35] Notably, the US market's total was only a little more than half of what was spent online for physical goods in China.

Table 7.1: e-Commerce: Country-wise Revenue Projections (US$ millions)

	2018	2023	Five-year Growth
China	$636.1	$1086.1	**70.7%**
United States	$504.6	$735.4	**45.7%**
France	$49.4	$71.9	**45.6%**
Australia	$18.6	$26.9	**44.6%**
Russia	$17.2	$24.8	**44.2%**
Canada	$39.9	$55.4	**38.8%**
Germany	$70.3	$95.3	**35.6%**
United Kingdom	$86.5	$113.6	**31.3%**
Japan	$81.7	$103.6	**26.8%**
South Korea	$63.7	$80.2	**25.9%**

Source: Statista

Amazon.com Inc. has emerged as the number one e-Commerce platform in the United States and is on the way to becoming a major e-Commerce giant across the globe.[36] "The retailer sells its own inventory and operates a marketplace on which other retailers sell their products through the Amazon platform, with Amazon taking a commission of each order".[37] Based on sales of retailers' owned products alone amounting to US$ 150 billion, Amazon was the clear global winner in 2017. Amazon accounted for as much as 43.1 per cent of all e-Commerce growth in the United States.

ii.) Emerging Technologies and e-Commerce

Technological developments are having an impact on the e-Commerce industry and are transforming the way consumers experience the online portal. Technologies are helping the customers to track their orders, find best products and deals etc. Following are the technologies that will impact the e-Commerce in future, the most:

Digital Supply Chain

An important part of e-Commerce industry is ensuring on-time delivery. Optimizing the supply chain management process is the key to ensure efficient and faster delivery. For better supply chain, the use of technology is inevitable. For example, for port management the use of autonomous freight shuttles are employed; for inventory management, hack proof RFID (Radio Frequency ID) tags are required; for warehouse management, Robots are being employed. Yet another example of use of emerging technology is deploying drones for delivery.[38]

Artificial Intelligence

One of the most important consideration for e-Commerce companies is customer experience. Many AI companies are transforming e-Commerce by developing unique solutions. For e-Commerce businesses, AI technology is becoming increasingly important "since it holds the potential for dynamic and adaptable targeting, which helps companies get the right pitch, at the right time, to the right customer, on the right platform."[39]

Payment Service

Faster, efficient and secure payment options have a very big role to play in the growth of e-Commerce industry. Gartner had predicted that 50 per cent consumers in mature markets will use smartphones or wearable-tech for making mobile payments by 2018.[40] According to Forrester's mobile payments forecast 2016-21, mobile payments in the EU-7 are estimated to rise at a Compound Annual Growth Rate (CAGR) of 19 per cent to reach US$ 148 billion by 2021. Forrester also predicts that in-person mobile payments will become a fast-growing segment, while remote mobile payments will constitute over 65 per cent of mobile payments by 2021.[41]

With the emergence of contactless payments options like Apple Pay and Google Wallet, security concerns have also emerged. The e-Commerce companies are trying to provide customers with payment options that are as secured and as smooth as possible.[42] It goes without saying that rather than banks, it would be FinTechs and TechFins that would be eminently suited to address the financial services involved in e-Commerce business worldwide.

C. 'Cryptocurrencies'

Cryptocurrency is simply a digital or virtual currency, which uses cryptography for its security. Technically, cryptocurrencies are decentralized systems based on the Blockchain Technology. One of the distinguishing feature that makes cryptocurrencies unique is lack of a central authority, which makes it insulated from any government intervention.

The Financial Action Task Force (FATF),[43] an intergovernmental group, provides a comprehensive formal definition of cryptocurrency: "The term 'cryptocurrency'..... refers to a math-based, decentralized convertible virtual currency that is protected by cryptography i.e., it incorporates principles of cryptography to implement a distributed decentralized, secure information economy".[44]

Cryptocurrency provides easier mode for transfer of funds directly between two parties in a transaction without the interference of the third party such as banks or the central banks. This anonymous nature of cryptocurrencies also makes them a host of illegal activities such as tax evasion, money laundering or terrorism financing.

According to Bank for International Settlements (BIS), cryptocurrency consists of three elements. First, a set of rules (the "protocol"), a computer code specifying how participants can transact. Secondly, a ledger storing the history of transactions. And thirdly, a "decentralized network of participants that update, store and read the ledger of transactions following the rules of the protocol".[45]

"With these elements, advocates claim, a cryptocurrency is not subject to the potentially misguided incentives of banks and sovereigns".[46] The first cryptocurrency was Bitcoin which is based on Blockchain Technology.

Cryptocurrency, in comparison to other digital money such as bank deposits have a distinguishing feature of digital peer-to-peer exchange. Digital banks have been around for decades but in contrast to their model, cryptocurrency transfers take place in decentralized circumstances without the presence of a central authority or medium to execute them.

Many economists are of the view that this currency should not be in the social system. Nobel Laureate Economist Joseph

Stiglitz believes that the cryptocurrency should be outlawed because it "doesn't serve any socially useful function" and that its popularity comes from the "potential for circumvention, lack of oversight".[47]

There have been many cryptocurrencies over the period of time. Bitcoin was the first of its kind. Apart from anonymity and the lack of control, there is also a serious concern regarding its highly instable value.[48] Illustratively, the pioneering cryptocurrency, Bitcoin, increased from a value below US$ 0.01 in 2009 to a high of US$19,783 on 17 December 2017. It almost tripled its value in one month in December 2017.[49] On December 1, 2017, there were a total of 19,430,457 Blockchain wallet users in existence. By May 11, 2018, this number increased by 27 per cent to 24,662,973 users. The rise in Blockchain wallets demonstrates the vast increase in cryptocurrency interest and usage.[50]

With the rapidly unfolding Banking-Finance 4.0 there is an imperative need for a new approach by central banks to regulate its evolution. These issues relating to regulatory and strategic development of Banking-Finance 4.0 are taken up in Chapter 13.

PART • THREE

NEW-AGE TECHNOLOGY: GLOBAL PUBLIC POLICY ISSUES

8

Technology: Right to Privacy (and Why it Matters)

"No one shall be subjected to arbitrary interference with his privacy, family, home or correspondence, nor to attacks upon his honour and reputation. Everyone has the right to the protection of the law against such interference on attacks"

The Universal Declaration of Human Rights (Article 12), United Nations, 1948

Introduction

The world has come a long way since the ratification of Article 12 of the UN's Universal Declaration of Human Rights. As discussed earlier (Chapter 5), the New-Age Technology underlying Industrial Revolution 4.0 promises to bring in enormous benefits for the society such as – from significantly improving productivity and economic growth to curing diseases and transforming education, etc. all over the world. But New-Age Technology would invariably entail costs. An especially worrisome aspect of its downsides have been the enormous societal cost in terms of sacrifice of individual privacy. A critical question that is being debated today is how can we promote innovation and reap the significantly large benefits of the Industrial Revolution 4.0, while limiting the potential harm to the society, especially the loss of privacy and individual autonomy and freedom.

Gary Daniels of American Civil Liberties Union (ACLU) once said: "We no longer have reason to fear if Big Brother will happen. We are right there in the middle of it now. Are we in danger of losing our right to privacy? That ship sailed a long time ago. We are all in a surveillance state. We don't have privacy".[1]

In the recent years, several scams have been exposed. Snowden disclosures in 2013 demonstrated how much information about people can end up in unexpected quarters. In 2017, the so called Equifax breach reportedly compromised identity information of around 146 million US citizens. The Facebook-Cambridge Analytica data controversy, in early 2018 allegedly revealed that personal data from profiles of millions of people was used for political advertising and propaganda purposes without their express consent.

Regrettably, there is no universally accepted privacy law among all countries. The imperative need for this has been recognized for quite some time by many including Bill Gates, who said: "Historically, privacy was almost implicit, because it was hard to find and gather information. But in the digital world, whether it is digital cameras or satellites or just what you click on, we need to have more explicit rules (on protection of privacy) – not just for government but for private companies (as well)".[2]

This chapter, which is focused on technology and privacy issues, has been organized as follows:

Section I discusses the meaning of privacy and its multitude dimensions. Section II dilates on privacy issues in the digital world. Against that backdrop, Section III presents a range of legal issues in the trade-off between technology and privacy. This is followed by Section IV which offers a cross-country perspective on privacy policy and concluding remarks.

I. Privacy: Meaning and Dimensions

Privacy has historical roots in discussions and debates among philosophers. Greek Philosopher Aristotle (384-322 BC) was probably the first one to make out a case for privacy. He made a distinction between public and private spheres, confining the authority of the government only to the public realm and not the private one.

This line of thought was carried forward, many centuries later, by British Political Economist and Thinker John Stuart Mill (1806-1873). In his famous book *On Liberty*, J.S. Mill emphasized the need to preserve a zone within which the liberty of citizen would be free from the authority of a nation-state. According to Mill, this could be done only by recognition of civil rights such as the individual right to privacy.[3]

Apparently, the first legal definition of privacy came from two American Jurists, Samuel Warren and Louis Brandeis. In 1890, they wrote an article titled, "The Right to Privacy", wherein privacy was defined as the "right to be let alone".[4] The two jurists recognized that sacred precincts of 'private' life have been invaded by the impact of (the then evolving) technology and that there is an imperative need to protect the "right to be let alone".[5]

The substance and limits of what may constitute as "private" varies across individuals, cultures and over time even in the same society. Not surprisingly, therefore, defining 'privacy' has been a challenging task.

As Professor Arthur R. Miller once stated privacy is "exasperatingly vague and evanescent".[6] American Moral Philosopher Judith Jarvis Thomson observed that, "the most striking thing about the right to privacy is that nobody seems to have any clear idea what it is".[7] Author Julie Inness related privacy with intimacy and pointed out that "privacy is linked with the common denominator of intimacy which covers intimate information, access and decisions".[8] However, this conception of privacy is too narrow since intimacy can be defined quite broadly and flexibly.

Several dimensions of privacy discussed in the legal literature have been reviewed below:

i.) Right to be Let Alone

Attributed to the US jurists Samuel Warren and Louis Brandeis, as mentioned above, this dimension has been interpreted to mean "the right of a person to choose seclusion from the attention of others if he or she wishes to do so, and the right to be immune from scrutiny or being observed in private settings, such as one's own home".[9] Although this early and somewhat vague concept did not describe privacy in a way that would help design privacy in broad legal terms, it did strengthen the notion of privacy rights for individuals and set in motion a legacy of discussion on those rights.[10]

ii.) Limited Access

Some theorists have visualized privacy as a system for limiting access to one's personal information. E.L. Godkin, for example, wrote way back in the year 1880 that "nothing is better worthy of legal protection than private life, or, in other words, the right of every man to keep his affair to himself, and to decide for himself to what extent they shall be the subject of public observation and discussion".[11]

iii.) Control over Information

According to Alan Westin, "privacy is the claim of individuals, groups, or institutions to determine for themselves when, how, and to what extent information about them is (to be) communicated to others."[12] In this dimension, privacy is seen essentially as the control people have over information about themselves.

iv.) Secrecy

Privacy is sometimes defined as an option for maintaining secrecy. According to Jurist Richard Posner, privacy is the right of people to "conceal information about themselves that others might use to their disadvantage".[13] In this dimension, privacy is seen as an eclectic sort of secrecy in which individuals tend to keep some information to themselves i.e. secret while choosing to make the remaining information public.[14]

v.) Self-Identity and Personal Growth

In yet another dimension, privacy is seen as a prerequisite for the development of a sense of self identity. The absence of barriers to privacy, in particular, is deemed to be antithetical in this process. According to Irwin Altman, privacy barriers "define and limit the boundaries of the self" and thus.... "serve to help define (the self)".[15]

Moreover, privacy may also be seen as a state that fosters personal growth, a process integral to the development of self-identity. It is argued that "without privacy – solitude, anonymity, and temporary releases from social roles – individuals would be unable to freely express themselves and to engage in self-discovery and self-criticism". Such self-discovery and self-criticism "contributes to one's understanding of oneself and shapes one's sense of identity".[16]

vi.) Intimacy

This dimension envisions privacy to be an essential part of the way people have intimate relationships with others. Since a part of human relationships involves individuals volunteering to self-disclose some information while withholding other, privacy can be seen as a part of the process through which human beings establish relationships with each other.

According to Philosopher James Rachels, who advanced this notion, privacy matters because "there is a close connection between our ability to control who has access to us and to information about us, and our ability to create and maintain different sorts of social relationships with different people".[17]

In sum, notwithstanding the difficulties in defining precisely, it is evident that the notion of privacy is closely associated, indeed intertwined with human dignity, freedom and independence of the individual.

II. Privacy Issues in the Digital World

As the world is progressing towards knowledge society, the underlying New-Age Technology has posed a formidable challenge to the notion of privacy.

All traditional norms of privacy and the implicit basic understanding that the content of individual communications should remain confidential, has gone haywire. Confidentiality is being eroded and threatened with the revolutionary advances in digital technology. Also, the fundamental principle that individuals ought to be able to control their personal data, prevent its collection by third parties and decide how it is used, can no longer be taken for granted.

As Marc Groman, former Senior Adviser for Privacy in the White House observed: This is "nearly impossible to implement in a world where personal data is collected, created, used, processed, analyzed, shared, transferred, copied, and stored in unprecedented ways and at an extraordinary speed and volume".[18] This includes the collection and creation of Big Data, i.e. personal data from an ever-increasing number of connected devices such as smartphones, smartwatches, connected cars, drones, personal assistants like Alexa or Siri, etc. This Internet of Things (IoT) – the network of physical objects, embedded with appropriate sensors, enable these objects to collect and exchange data. "The network of 'objects' connected is expected to expand from an estimated 8.4 billion devices in 2017 to a projected 20.4 billion in 2020".[19]

Today, more data about human beings is being generated, faster and faster and from larger number of devices than ever before. And, there is no option to opt out of the emerging data-intensive world, simply because the use of the emerging New-Age Technology and sharing personal information by people is increasingly becoming almost mandatory for participation in modern society. Internet access and use of the New-Age Technology is increasingly becoming necessary for education, employment and, indeed, for full participation in civic and economic life.

Privacy is being threatened and undermined in this transition process, day in and day out, wherever people go, without them realizing it. An article in iNews points out, among other things, the following possibilities:[20]

i.) In the so-called 'smart' cities, anyone from municipal authorities to digital start-ups could absorb personal data as people move around and interact in public spaces. This data gathering can take place easily without their knowledge, through their mobile phones and sensors possibly embedded in everything including the traffic lights.

ii.) In 'rogue' wifi hotspots (with misleading names such as "stirbucks-wifi") logging into unsecure networks often enable cybercriminals to eavesdrop on what people do online.

iii.) Mass surveillance programs run by intelligence agencies (as well as police) often deploy a wide range of highly intrusive devices. Increasing use of facial recognition technology and body-worn cameras mean that people's ability to be anonymous is fast disappearing.

iv.) Checking social media posts and photographs is yet another way of encroaching upon people's privacy.

How has all this become possible, now? The answer, of course, lies in the emerging New-Age Technology. There are at least three major technological developments that are deeply affecting privacy:[21]

- The increase in data creation and the subsequent collection of vast amounts of personal data,
- Globalization of data market and the ability of anyone to collate and examine data, and
- Lack of control mechanisms for digital data, which have existed for long to protect analog data.

The entire emerging technology contributes to this invasion of privacy. Especially noteworthy are technological advances in respect of (a) Robots (b) Data Analytics, Artificial Intelligence (AI) and Machine Learning (ML) and (c) Internet of Things (IoT).

(a) Robots

Robots, by definition, is a "machine that is equipped with the ability to sense, process and record the space around it".[22] Robots are "capable of doing things that humans cannot do, they are capable of going places where humans cannot go and seeing things that humans cannot see".[23] Given their sizes and shapes, equipped with sophisticated sensors and processors, Robots magnify the way in which humans observe.[24] This is what makes them reliable and versatile.[25]

Depending on their range of work, Robots can be differentiated into different categories such as Industrial Robots and fully Autonomous Robots, the latter are capable of functioning without any human control. These Autonomous Robots are "fully AI-integrated entities, capable of utilizing all the information or data available to them."[26]

One of the major concerns with Robots ecosystem is that Robots have access to a wealth of information. A Personal Robot, for example, by its very design, will have access to a wide range of information of its users ranging from health to financial information.[27] "There is a strong possibility that the sensitive information accessed by the Robot may be disclosed to any third party, either intentionally or accidentally".[28]

M. Ryan Calo has articulated the way in which Robots intrude on privacy.[29] First, they greatly facilitate in direct surveillance. Secondly, they introduce new access points for third parties – from the government, to hackers, to hitherto protected spaces such as the interiors of homes. Finally, Robots in their interaction with human beings, get plenty of opportunities to gain a wealth of information from them. Moreover, a Robot's "tireless memory along with its capability of not being affected by humanly emotions and feelings, such as embarrassment, gives them advantage over humans".[30]

As mentioned above, the disclosure of information by the Robots can be either intentional or accidental. "The intentional disclosure of information to the third party can be engineered by either the developers of the Robot or its users".[31] On the other hand, an accidental disclosure of the information implies a disclosure where the computer system to which the Robot is connected, is infected by malware i.e. it is hacked.[32] In both cases, there is not only a loss of privacy but also a breach of security. A study carried out on Personal Robots as a part of the European Union (EU) integrated project COGNIRON, concluded that "a Personal Robot may be like a 'walking CCTV camera' that is capable of recording the information about its environment with the potential threat of autonomously revealing this wealth of information to third parties".[33]

Thus, a Robot whether controlled manually or autonomously, possesses the capability of entering and breaching the private spaces of an individual that typically are exclusive and strictly personal in nature. Given the enormous quantities of data that is readily available to a Robot during the course of its working, it makes them a threat to privacy and security.

(b) Data Analytics, Artificial Intelligence and Machine Learning

The potential effects of Artificial Intelligence (AI) are likely to be far-reaching and the technology is all set to get integrated into almost all human endeavours. As AI evolves, it may be possible to predict and even manipulate the behavior of different strata of the society.

Recall the basics of AI (Chapter 2) again for a moment. While learning by the inductive method, humans collect a large amount of data. They try to interpret that data to draw patterns of behavior in whatever they are observing and then draw conclusions. The larger the datasets, the more accurate the conclusions are. The same is true for Machine Learning (ML) as well; the only difference being that machines can handle much larger and exponentially growing amounts of data and due to this, their conclusions are likely to be more accurate. In this post Digital Revolution era, there is already 1.2 million terabytes of data (one terabyte is 1000 gigabytes) on the Internet[34] and it is not humanly possible to assess such huge amount of data. However, data is something that AI thrives on. More the data, better are its conclusions.

Interestingly, a Stanford University study (2018) found that computers are better judges of personality than friends and family. The Study tried to determine whether a machine learning algorithm can understand a person's personality better than friends and family based on the content that they like on Facebook. It was found that the algorithm requires only ten likes to predict a person's behavior as accurately as his or her colleague; just a hundred likes for predicting with accuracy as much as a family member and only two hundred likes to predict behavior as accurately as his or her spouse.[35]

Importantly, it may be noted that the emergence of AI, Machine Learning (ML) and the Big Data have underscored the need for the "Right to Privacy" in order to have free and fair elections. This is because the threat to democracy arises when "Right to Privacy" is violated. This was clearly demonstrated in the Cambridge Analytica case in which, allegedly, large data relating to political views of people was analyzed and their views were influenced over the Internet. That, in turn, reportedly, influenced the results of the 2016 American Presidential elections. Needless to say that the issue of data privacy goes far beyond the question of why a machine should be knowing more about a person than his or her spouse!

(c) Internet of Things (IoT)

Suppose that one never sets foot in a public space again and, one closes all personal social media accounts. Surely, people can have some privacy within their own home, right?

Not really. The Internet of Things (IoT) i.e. Internet-connected devices "ranging from Smart TVs to virtual assistants like Siri and Alexa, while offering convenience and efficiency, "are also building a detailed, revealing and intimate data profiles of people, which manufactures and government surveillance programmes can tap into".[36]

Given this constant intrusion on people's privacy, legal issues involved in protection of privacy assume special significance and are as discussed next.

III. Legal Issues in Technology vs. Privacy

The tradeoff between technological advancements and concerns for privacy have given rise to a whole range of legal issues. These are discussed below in the context of different domains.

(a) Data Privacy

Data Privacy essentially refers to the relationship between the emerging technology and the legal right to collect and share data.

Privacy concerns arise wherever "uniquely identifiable data" relating to a person or persons is collected and stored, in digital form or otherwise. In some cases, there are concerns about how data itself is collected and stored, or who has access to that information. Other legal issues include whether an individual has any ownership rights over data about him or her, and whether there is a right to view, verify, and challenge that information.

(b) Financial Privacy

Financial Privacy refers to financial information of individuals. Financial Privacy especially includes confidentiality regarding the bank accounts of individuals and information about the bank where the account is opened. Guarding a person's financial privacy is obviously important for avoiding fraud(s), including identity theft. Information about a person's purchases can reveal a great deal about his or her preferences, places the person has visited, his or her contacts, products (such as medications) they use, their activities and habits, etc. which could be misused if available to unscrupulous third parties.

(c) Internet Privacy

Internet Privacy is all about the ability of a person to determine what information he or she reveals or withholds about self-online, who has access to this information, and the purposes for which one's information may or may not be used. For example, Internet users may be concerned that many of the websites which they visit collect, store, and possibly share "uniquely identifiable information" about them. Similarly, Internet email users consider their emails to be private and hence would be concerned if their emails were being accessed, read, stored or forwarded to third parties without their consent.

The Internet and the emerging technology have brought new concerns about privacy in view of the fact that computers can permanently store records of everything: where every online photo, status update, Twitter post and blog entry, by and about people, can be stored forever.

(d) Medical Privacy

Medical Privacy refers to protected health information. It allows a person to withhold his or her medical records and related information from others. One can think of several motives behind this, from fears that it might affect their insurance coverage or employment, or simply avoidance of the possible embarrassment of disclosing medical conditions or treatments. Medical information could also inform other aspects of one's personal life, such as sexual preferences or proclivity, which people may not want to reveal.

(e) Corporate Privacy

Corporate Privacy refers to the privacy rights of corporate entities and their officials, especially senior executives of large corporations. Motivations for corporate privacy can frequently arise out of competition and also in the context of obligations for public disclosures under securities and corporate laws of the land relevant nation-state.

(f) Organizational Privacy

Government agencies, groups/societies and other organizations may desire to keep their activities (or confidential matters) from being revealed to other organizations or individuals. They may adopt various security practices and controls in order to keep private information under wraps. Organizations may seek legal protection for their confidential matters. Illustratively, a government administration may be able to invoke executive privilege or declare certain information as classified.

The emerging technology-enabled and greatly enhanced ability to gather and send information has had seriously adverse implications for retaining privacy. With large scale information systems becoming increasingly common, enormous information is going to be stored in many databases worldwide. An individual typically, has no way of knowing or controlling all the information about him or her that others may hold or have access to. Such information could potentially be sold to others for profit and/or be used without the knowledge or approval of the individual concerned. That is precisely why, in this time and age, privacy is assuming an over-riding importance.

The concept of privacy is becoming highly significant as more systems, controlling larger information, appear. The consequences of privacy violations are also likely to be more severe than ever before. Not surprisingly, the international community is struggling to keep pace with the emerging technology and at the same time, adopting appropriate privacy laws in respective nation-states.

IV. Privacy Policy: A Cross-country Perspective

Most countries provide the rights to privacy for their citizen in their Constitutions. For example, Constitution of Brazil says "the privacy, private life, honor and image of people are inviolable"; the Constitution of South Africa says that "everyone has a right to privacy"; and the Constitution of the Republic of Korea says "the privacy of no citizen shall be infringed".[37] "Several countries whose Constitutions do not explicitly describe privacy rights, their respective Court decisions "have interpreted their Constitutions to intend to give privacy rights".[38] On the other hand, many countries have broad privacy laws outside their Constitutions; these include Australia's Privacy Act 1988, Argentina's Law for the Protection of Personal Data of 2000, Canada's Personal Information Protection and Electronic Documents Act, 2000 and Japan's Personal Information Protection Law, 2003.[39]

Beyond national privacy laws, of course, there are regional and international privacy agreements. As mentioned earlier, the United Nations Universal Declaration of Human Rights says "No one shall be subjected to arbitrary interference with his privacy, family, home or correspondence, nor to attacks upon his honor and reputation". The Organization for Economic Co-operation and Development (OECD) published its Privacy Guidelines in 1980. The European Union's General Data Protection Regulation (GDPR), 2016 (operationalized in 2018) guides privacy protection in Europe. Likewise, the 2004 Privacy Framework by the Asia-Pacific Economic Cooperation (APEC) is a privacy protection agreement for the members of that organization.

(a) Approaches to Privacy Policy

Approaches to privacy policy can broadly be divided into two categories: free market and consumer protection.[40] In the Free Market approach, commercial entities are generally allowed to do what they wish, with the presumption that consumers will choose to do business with only those corporations that respect their privacy. If some companies do not show adequate respect of privacy, they will lose their market share and even be forced out of business. A shortcoming of this approach is that claims of privacy protection made by companies may be difficult for consumers to verify *a priori*. A practical example of the free market approach is voluntary OECD Guidelines on the Protection of Privacy and Transborder Flows of Personal Data.

In contrast, under the Consumer Protection approach, the basic premise is that individuals might not have the knowledge to make informed choices, or might not have reasonable alternatives. This is so because, as shown by some researchers, "most privacy policies are above the reading level of the average person".[41] Accordingly, this approach advocates greater government involvement in definition and enforcement of privacy standards.

(b) General Data Protection Regulation (GDPR)

With a view to address the privacy issues, the European Union (EU) set up a legal framework for issuing guidelines for the collection and processing of personal information of individuals within its area of influence. The General Data Protection Regulation (GDPR), as it is called, became operational with effect from May 25, 2018.

Article 5 of the EU-GDPR mandates that processing of personal data must be done in a transparent manner. As noted before (Chapter 2), it is impossible to determine the way Deep Learning algorithms come to particular conclusion. In light of Article 5, since Deep Learning algorithms are not capable of being transparent, they may not altogether be allowed under the GDPR. A related issue is that companies may not be comfortable in revealing the algorithms they are using in order to process personal data.

Article 6 of the GDPR mandates that while collecting personal data, consent must be sought from the data subject. This also may be a cumbersome exercise as Machine Learning involves collection of huge amounts of data.

The GDPR also mandates the principle of ***Privacy by Design***, which essentially stipulates that Artificial Intelligence (AI) must be designed in such a manner that it would only collect data necessary to fulfill the purpose for which it was created. This limits the horizon of research in AI technologies only to narrow AI (i.e. the type of AI that is limited only to a specific purpose such as self-driving cars, whereas General Purpose AI is closer to what a human being is like and can perform several functions).[42]

USA

In the United States, there is no express provision with reference to Right to Privacy, though implicitly it is recognized.[43] Unlike France, the US history is not the one where the Judiciary built an impressive jurisprudence around privacy upon a Roman Foundation in the 19th-century.[44] It is also not like the UK, where privacy is an unrecognized concept even today.[45] It's a *sui generis* creation in the US.[46]

In the US, conceptualization of privacy revolves around the subject of personality rights. Today "right to privacy is a gateway to protect against unauthorized use of an individual's name, likeness, publicity rights, confidences, compositions, life history – an expression of autonomous decision making".[47]

UK

In the United Kingdom, it was only in the year 2000 that the Human Rights Act came into force, which incorporated the European Convention of Human Rights (Article 8), which protects the right to private and family life. Until then, it was just a piecemeal protection in the area of privacy consisting of a mixture of statute and common law.

Presently, there is an ongoing debate regarding the UK government's "Snoopers Charter", which retains the communications data of everyone for a set period. If this makes way in the Statute book, that would be a step far away from privacy in the name of balancing security.[48]

These policies, legislations and judicial decisions are symptomatic of the manner in which "the Right to Privacy is being construed in diverse jurisdictions based on the histories of the societies they govern and the perceived challenges before them".[49]

Autonomy is so intrinsic to our lives that it has been hailed recently as a fundamental right in India and comes within the "right to life and liberty".[50] Therefore, it must be recognized that the attitude of ignoring privacy effectively amounts to undermining the democratic structure of a nation-state.

9

Technology: Social Media and Democratic Governance

"Social media is something of a double-edged sword. At its best, social media offers unprecedented opportunities for marginalized people to speak and bring much needed attention to the issues they face. At its worst, social media also offers 'everyone' an unprecedented opportunity to share in collective outrage without reflection".

Roxane Gay[1]

Introduction

Social media has really come a long way in a rather short span of time. Regrettably, the change has not been necessarily for the better. A young actor, Lazzie David reportedly said: "It is so funny how social media was just this fun thing, and now it's this monster that consumes so many millennial lives". Another young model, Kendall Jenner in fact, goes to the extent of saying: "I think social media has taken over lives, and it's kind of sad".[2]

This chapter, focuses on technology, social media and democratic governance is organized into two sections. Section I discusses the major issues of democratic governance raised by the emergence and proliferation of social media, and Section II dilates on surveillance issues in the context of social media.

I. Social Media and Democratic Governance

Social media is essentially, a set of Internet-based interactive technologies that facilitate creation and sharing of data (such as text posts, comments, digital photos, videos and other forms of expression) through virtual communities and networks. In fact, users create their own profiles and social-media enables development of online social networks by connecting each user's profile with those of other individuals and groups.[3]

Social media is a new powerful forum that helps people to exchange ideas, connect with each other, seek advice, offer guidance and relate to (and possibly mobilize) for a cause. It has completely eliminated the communication barriers and created a totally decentralized communication channel and, in the process, has opened a gateway for people to have a voice and to participate in a functioning democracy.

Today, across the world, social media has been influencing societies, economies, people's attitudes, behaviors, cultures, and even their world view.

In recent years, there has been a phenomenal growth of social media platforms, especially as apps on mobile devices such as smartphones. Six apps have more than a billion users each.[4]

These include:

Facebook	2.32	billion users
YouTube	1.90	billion users
WhatsApp	1.60	billion users
Facebook Messenger	1.30	billion users
WeChat	1.1	billion users
Instagram	1.0	billion users

The total number of social media users worldwide, which was under one billion (0.97 billion) in 2010, rose to 2.77 billion in 2019 and by the year 2021, the number is estimated to exceed the 3 billion mark. (It may be noted that the global population in 2019 is estimated to be around 7.7 billion.) In a country like India, which is the second largest online market (next only to China), the number of social network users in 2019 was placed at 351 million and is projected to grow to around 423 million by 2022.[5]

Not too long ago, the Internet and social media was seen as a great source of human empowerment. It was seen as a potent instrument for correcting the power asymmetry between the individual and large organizations (including the government). Many hoped – even were convinced that the Internet (and social media) would redress the growing inequality of influence and give more power to the individual. Seemingly, the ability to influence and shape opinion was no longer the monopoly of mass-media organizations or the government. "Devolution and decentralization of power seemed inevitable and (in fact) just around the corner".[6]

Regrettably, however, that perceived future has not become a reality, at least, as yet. "The proliferation of a multitude of sources made for an overload and led quickly to the evolution of content aggregators. These portals are (not only) gateways to content, but also gatekeepers. This makes the concept of a completely democratized Internet as unrealistic one".[7]

Earlier it was believed that social media would facilitate the dialogue among people by providing a platform for exchanging differing viewpoints and perspectives. Unfortunately, quite the opposite seems to be happening in a number of ways:

1. The biggest strength of social media of course, has been the ability of spreading information to millions of people in a matter of seconds. This would certainly be a great advantage in the event of an emergency. On the other hand, implications would be disastrous if unsubstantiated reports are passed on. For example, fake news or rumours spread to thousands of people or more through the so-called 'forwards'.

Typically, the more sensational the 'news' post, the faster and wider it gets circulated. The impact multiplier here is the so called 'echo chambers effect' (defined as a situation where beliefs are amplified or re-enforced by communication and repetition inside a closed system).The 'echo chamber effect' has, directly or indirectly, promoted the planting of fake news which, more often than not, has led to fear, enmity and hate, besides occasionally, even triggering riots.

A classic oft repeated example is that on April 2013, the Syrian Electronic Army attacked the *Associated Press* (AP) Twitter account and sent a message to millions of readers of the AP's Twitter that said "Breaking News: Two explosions in the White House and Barack Obama is injured". The news was of course, false. But the stock market plunged within minutes; the New York Stock Exchange dropped US$ 136 billion in market valuation.[8] This is nothing short of an act of Cyberterrorism that can destabilize a nation's economy and undermine security besides creating a global turmoil.

2. Social media "mining" is a technique of analyzing data to detect patterns. It is a process of representing, analyzing, and extracting actionable patterns from data collected from people's activities on social media. Illustratively, Google mines data in many ways including using an algorithm in Gmail to analyze information. This use of information will then indicate to them the type of advertisements to be shown to the users when they use Gmail. Facebook also has partnered with many data mining companies to use customer information for 'targeted advertising'.[9]

Big Data Analytics and Artificial Intelligence (AI) could be used by platforms like Google, not only for targeted advertising, but also for 'customizing' news.

Reading news—valid or fake—that interests the viewer and is consistent with his or her beliefs and the world view, obviously, amplifies the 'echo chamber effect', further.

Twitter is increasingly being targeted by big-time marketers. For gaining larger number of followers, marketers use advanced scripts and manipulation techniques that distort and abuse the very idea of social media. British-American entrepreneur and author Andrew Keen commented on social media thus, "Out of this anarchy, it suddenly became clear that what was governing the infinite monkeys now inputting away on the Internet was the law of digital Darwinism, the survival of the loudest and most opinionated. Under these rules, the only way to intellectually prevail is by infinite filibustering."[10]

Rather than promoting the dialogue-based cross-cultural understanding, seemingly, what social media has ended up doing is reinforcing sectarian identity and sharpening the societal divisions. As Karnik (2018) puts it, "the hope that these technologies would democratize, empower and enlighten seems to have been thwarted. May be it was a false dawn, after all".[11]

In the prevailing environment, the world over, characterized by ultra-nationalism and socio-cultural schisms, social media seems to be adding fuel to the fire by exacerbating the dangers of polarization.

3. As far as politics is concerned, social media has a range of uses in political processes and activities. Social media was once hailed, as a champion of democracy since it enabled anyone with an Internet connection to become a content creator and thereby empowering the users. This role of social media in democratizing media participation, and ushering in a new era of participatory democracy where all users are able to contribute news and comments, has fallen grossly short of the ideals. "Online media audience

members are largely passive consumers, while content creation is dominated by a small number of users who post comments and write new content".[12]

According to the Pew Research Center, a majority of Americans, at least occasionally, receive news from social media. As a result of algorithms on social media that filter and display news content that is likely to match the users' political preferences, receiving such news from social media could mean an increase in political polarization due to selective exposure. Selective exposure occurs when an individual favours information which supports his or her beliefs and avoids those details which conflict with their beliefs. A study by Hayat and Samuel-Azran conducted during the 2016 US Presidential election observed an 'echo chamber effect' effect of selective exposure among 27,811 Twitter users following the content of cable news shows. The Twitter users observed in the study were found to have little interaction with users and content whose beliefs were different from their own, possibly heightening polarization effects.

4. Yet another, rather dangerous aspect of social media is its rapid adoption by terrorist groups. Social media comes in handy for terrorists for gathering information, recruiting members, fund raising and propaganda schemes.[13]

   As explained in Weimann's report, social media allows terrorists to interact with each other and therefore, enables them to get new recruits. Terrorists know that people accessing social media, are generally impressionable young people, the perfect target groups for terrorists. Recruiting new members has become easier now with the growing use of social media, especially, through the new movement called "lone-wolf terrorism". "Social media is helping …… (this kind of) terrorism in which there is somebody who trains, guides and launches them….. lone wolves were never truly alone".[14]

In sum, social media, notwithstanding its great benefits, is increasingly becoming a tool for individual criminals and terrorist groups. It is thus, difficult to say that social media is positively contributing towards democratic governance. Like any other technology, it is a double-edged weapon, after all.

II. Surveillance of Social Media

In several developed countries, programs intended to spy over individuals online in the name of national security and public interest have been in force for quite some time. For instance, the UK police use social media intelligence technology popularly known as SOCMINT. This involves snooping on content posted to public or private groups or pages. It also means grabbing all data from a social media platform including contents of posts and data about one's behavior (such as his or her likes and shares) with a view to generate predictions about users.[15] In addition to this, police officers can even create fake profiles, covertly adding users as "friends" so as to gain access to information that is otherwise private or use facial recognition technology to track and pinpoint one's whereabouts.[16]

Collecting the data on activities of civilians through social media has numerous implications. The Big Data collected can be used by the authority to curtail freedom of speech, make predictive analysis and thereby influence the public favorably. There is also the risk of being hacked illegally, for financial and other motivations.[17]

Facebook-Cambridge Analytica Data controversy (March 2018) is one such example of how data collected from social media can be allegedly misused by private players and at times even by the State authorities. The UK-based political consulting firm Cambridge Analytica was alleged to have mined the data of more than 50 million individuals from social networking sites without informing them or taking their consent in order to use it for political campaigns.[18] There were allegations that the company acquired access to Facebook profiles, locations and friends' data of individuals by violating social media policies. Reportedly, this was done for the purpose of target messaging in order to influence the voting patterns of the people for Brexit and for 2016 US Presidential elections.[19]

The Data Scandal erupted in March 2018 when a whistleblower, an ex-Cambridge Analytica employee, Christopher Wyllie, made the revelation.[20] It was only after that a serious public discussion on ethical standards of social media companies and political organizations, and the need for a stronger consumer protection laws in online media took center-stage.

These surveillance programs conducted by national and state governments, and running under the aegis of different security agencies and tech policy companies, operate with the stated aim of preservation of public peace and tranquility and curtailment of any elements of public disturbance.[21] However, it is important to realize that in the absence of any effective regulation to govern the operation of these applications, it is exceedingly difficult to determine whether or not the basic human and Constitutional rights of free speech, expression, and privacy of individuals are curtailed under the façade of community peace and interest of general public.[22]

Not surprisingly, in May 2018, the European Union put in place the General Data Protection Regulation (GDPR), one of the most stringent data protection laws, to protect the personal information of individuals. The rights include "making it easier for people to access the data companies that hold information about them, a new fines regime and a clear responsibility for organization to obtain the consent of the people they collect information about".[23]

10

Technology and Issues in Cybersecurity

"The diverse threats we face are increasingly cyber-based. Much of America's most sensitive data is stored on computers. We are losing data, money, and ideas through cyber intrusions. This threatens innovation and, as citizens, we are also increasingly vulnerable to losing our personal information...

We face cyber threats from state-sponsored hackers, hackers on hire, global cyber syndicates and terrorists. They steal our State secrets, our trade secrets, our technology and our ideas– things of incredible value to all of us."

James Comey[1]

Former Director, US Federal Bureau of Investigation (FBI)

Introduction

Cybersecurity has become the most important dimension of security globally. This phenomenon can be attributed, almost entirely, to the emergence of the New-Age Technology that underlies the Industrial Revolution 4.0.

In today's world, the Internet serves as the most valued treasure trove of information and mechanism for social interactions. The expansion of the Internet along with the emergence of the New-Age revolutionary technology has brought not only increased efficiency but has also introduced new modes of communication, social interactions, and information sharing. This online space of social interaction is Cyberspace. It is essentially the environment in which communications over computer networks occur within and across territorial boundaries. It is nothing but a global interactive domain consisting of digital networks which are used to store, modify and share information.

Almost everybody is connected to the Cyberspace one way or the other; the governments for collecting and storing data related to citizens, educational and medical institutions for communicating or maintaining the records of students and patients respectively and even common people who store data on their smartphones.

The entity with whom this data is entrusted (may it be the government or a private corporation) is expected to be responsible for its protection, which leads to the concept of Cybersecurity. More formally, Cybersecurity is defined as "the collection of tools, policies, security concepts, security safeguards, guidelines, risk management approaches, actions, training, best practices, assurance, and technologies that can be used to protect the cyber environment and organization and user's assets".[2] Organization and user's assets in this context typically include "connected computing devices, personnel, infrastructure, applications, services, telecommunications systems, and the totality of transmitted and/or stored information in the cyber environment".[3] In other words, Cybersecurity essentially refers to the techniques of protecting computer networks, programs and data from unauthorized access or attacks. It encompasses a whole range of threats to digital infrastructure, means to protect digital information and the systems it depends upon for communication.

According to the Global Risks Report 2019, "concerns about data fraud and Cyberattacks were prominent again in the GPRS".[4] The Report also highlighted a number of other technological vulnerabilities. There "were further massive data breaches in 2018, new hardware weaknesses were revealed, and research pointed to the potential uses of artificial intelligence to engineer more potent cyberattacks".[5]

The chapter focusses on the nexus between the emerging New-Age Technology and the growing concerns regarding Cybersecurity.

The chapter is organized as follows:

Section I explains how vulnerabilities in Cyberspace arise from the emerging technologies. Section II discusses "where" and "how" the vulnerabilities occur in a nation-state. Against that background, Section III dilates on some crucial dimensions of Cybersecurity i.e. Cybercrimes, Cyberespionage, Cyberterrorism and Cyberwarfare, in that order. The final section, i.e. Section IV offers a cross-country perspective on Cybersecurity in the context of the framework laid down by the United Nations.

I. Technology and Vulnerabilities in Cyberspace

Vulnerabilities in Cyberspace arising from the New-Age Technology may be illustrated with reference to (a) Information and Communication Technology (ICT) and Cloud Computing, (b) Big Data and Internet and (c) Internet of Things (IoT).

A. Vulnerability Arising from ICT and Cloud Computing

Information and Communication Technology (ICT) is an ever evolving and ever-expanding sector. Each passing day brings with it more technological advancements and newer modes of communication. In an era, where individuals rely heavily on the Internet and spend significant amounts of time with computers, securing Cyberspace and the communication sector becomes an over-riding priority.

Moreover, Cloud services (accessing computer, different software applications etc. through the Internet from any remote location) have blurred the institutional, organizational as well as national boundaries by enabling free flow of data from any distant, hard to reach location. The emergence of these technologies, therefore, raise serious concerns for Cybersecurity.

B. Vulnerability Arising from Big Data and Internet

Big Data is not only the characteristic of the emerging New-Age Technologies, it is the new focal point for Cybersecurity as well. If a nation-state does not pay sufficient attention to the storage, protection, and utilization of Big Data, valuable information is lost. More importantly, the damage, alteration and leakage of information can be hazardous to the country's information security.

One of the main concerns for any data protection regime is this: while the web is a source of lawful activity – both personal and commercial – concerns of national security arise nevertheless, simply because the seamless functioning of the web can be exploited by terrorists to wreak havoc and destroy parts of civilized societies.

As Richard Posner, an American Jurist and Economist noted:[6]

> Privacy is the terrorist's best friend, and the terrorist's privacy has been enhanced by the same technological developments that have both made data mining feasible and elicited vast quantities of personal information from innocents: the internet, with its anonymity, and the secure encryption of digitized data which, when combined with that anonymity, make the internet a powerful tool of conspiracy. The government has a compelling need to exploit digitization in defense of national security.

Every nation-state needs to clearly understand that distributed, virtual and service-oriented technologies in Cloud Computing capability constitute its technical foundation. Attacks on and breakdowns of Cloud Computing platforms could cause massive server paralysis with disastrous implications.

C. Vulnerability Arising from the Internet of Things (IoT)

The security of the Internet of Things (IoT) cannot be taken for granted either. The Internet-of-Things has numerous nodes, which are not easy to manage. Data stored at the nodes may be altered or forged, and this could be a major Cybersecurity challenge.

For example, the hackers may launch attacks using smart TVs, refrigerators, wireless loudspeakers and even medical devices.

In the world's first attack against the Internet-of-Things, over 100,000 Internet-connected "intelligent" home appliances were connected into a vicious network under the control of hackers. Immediately thereafter about 750,000 phishing emails (seeking sensitive information such as usernames, passwords and credit details) were sent within two weeks in the disguise of a trustworthy entity to unguarded victims.[7]

II. Cyberattacks: Where and How

Vulnerability in the Cyberspace often engenders Cyberattacks. While in principle, Cyberattacks can hit any sector of the economy and society, attention is focused here on:

- Critical Information Infrastructure (CII)
- Defense Sector
- Healthcare Sector, and
- Finance Sector

A. Cyberattacks on CII

According to the Global Risk Report 2019, the year 2018 "provided further evidence that cyberattacks pose risks to critical infrastructure, prompting countries to strengthen their screening of cross-border partnerships on national security grounds".[8]

Critical Information Infrastructure (CII) has generally been defined as "those facilities, systems or functions whose incapacity or destruction would cause a debilitating impact on national security, governance, economy and social well-being of a nation".[9]

A classic example of an attack on CII is Stuxnet – a malware that was used in 2011 for causing substantial damage to the centrifuges used in an Iranian nuclear facility. Yet another case is that of shutting down of a Ukrainian regional electricity distribution company. The company reported service outages to customers due to a third party's illegal entry into its computer and Supervisory Control and Data Acquisition (SCADA) systems. Seven substations were hit, leaving about 225,000 customers without power for three hours. Ukrainian news media reported that a foreign attacker remotely controlled the SCADA distribution management system to cause the outage.[10]

The Cyberattacks on the CII can cause huge financial losses, environmental damage and pose a threat to defense, security and integrity of the country.[11] A 2003 study found that disproportionately large number of global terrorist activities, post-2001 attack, were directed towards CII.[12]

It is increasingly being realized by the policymakers and governments world over that any country can be attacked remotely as well.[13] Such attacks can cause widespread damage to telephone networks and even bring down aircraft flight control systems. A 2008 incident in Poland is a case in point, when a 14-year-old took control of the city's tram, derailed vehicles and caused injuries. These occurrences can be prevented only by building strong firewall systems or intrusion detection systems.

B. Defense Sector

"War" said Napolean Bonaparte, "is ninety per cent information". This is more relevant today than ever before. Traditionally, the defense sector had been subjected to conventional threats until revolutionary advances were made in the field of information technology (IT).This has brought about a dramatic transformation in the defense sector, wherein, the Cyberspace has now emerged as the fifth domain in addition to land, sea, air and space.

The emerging revolutionary technology is used these days in upgradation of real-time intelligence and weapons, surveillance and reconnaissance systems. It is further used in collecting and storing volumes of classified data critical for national as well as global security. While these developments have enabled the defense sector to function more efficiently, they also have made it vulnerable to hazardous Cyberthreats. Vulnerabilities attract cyber-attacks and hackers succeed in exploiting these weaknesses and penetrate into protected computer systems of nations.

Hackers in the arena of Cybersecurity have the potential to access multiple weapon system designs and indulge in theft of sensitive intellectual property. What is even more worrisome is that hackers no longer require sophisticated weaponry, instead can successfully conduct dangerous attacks on security through low cost computing devices. Moreover, Cyberattacks don't require trained forces, vehicles, weapons or military bases to reach their targets. Once attacked, it is exceedingly difficult to punish the attackers since the digital footprints are mostly well-covered, making it nearly impossible to trace them. This makes retaliation difficult, often forcing governments to take decisions, which are not well-informed.

C. Healthcare Sector

Traditionally, it was believed that protective measures against Cyber-threats in the field of medical science is not required. In fact, there was no association between healthcare and Cyber-security. There is a growing reliance on the use of technology to provide effective healthcare, especially for chronic conditions. The emerging technologies can provide storage for health records, monitor health and provide medication. As healthcare devices continue to develop, so does their connectivity. Devices in hospitals are now interconnected so as to ensure efficiency, automation and remote monitoring.

Regrettably, this is changing now. The technology used for medical devices is now being used by hackers as an entry point to bypass the firewalls of hospital systems. These systems store millions of valuable patients' data. A stolen medical data is often more valuable than a financial one as the former is considered to be extremely confidential. Moreover, healthcare systems not only store data related to patients' health, but also store bank account and credit/debit card details used to make medical payments. This results in incidents of fraud. The data can be used to open bank accounts, take loans and even secure passports. The latter in turn, can easily be used for organized crimes and terrorism. This is most unfortunate because Cybersecurity in healthcare is often not accorded the priority it deserves, resulting in underdeveloped security expertise. Weak and outdated protection of the healthcare systems not only result in financial bonanza for hackers but also threatens human lives.

Globally, the healthcare industry has faced several Cyber-attacks. The WannaCry ransomware attack of May 2017 crippled the healthcare industry worldwide, holding life-saving information hostage, and demonstrating that risk to Cybersecurity is a threat not to be taken lightly. Reportedly the WannaCry attacks affected more than 200,000 computers across 150 countries with total damages running into millions of US dollars. By the end of the year, the US, UK and Australia formally indicated that North Korea was behind the attack.[14]

The only way to address the Cybersecurity challenge in the healthcare system is maintaining cyber resistance. It is necessary for medical systems and devices to create regular backups of data so as to ensure a quick recovery in the event of a Cyberattack. Furthermore, it is important that the softwares are up to date. Majority of the softwares used even in the most advanced hospitals are outdated and are often not compatible with new and developed anti-virus and anti-theft applications.

This, of course, calls for adequate budgetary provisions for investments and incentives to invest in the health sector. Only an increased fund flow, as well as enhanced awareness in the health care sector is likely to ensure a safe cyber environment in this critical sector.

D. Financial Sector

The finance sector has seen numerous challenges to Cybersecurity over the past few years, raising concerns all over the world.

Emerging technologies and waves of digitization have brought in their wake, new challenges and exposed organizations to new risks. It is estimated that Cyberattacks cost companies approximately US$500 billion in damages every year.[15]

According to 2019 Internet Security Threat Report[16], cyber criminals are getting rich quick with the so-called "form jacking" (i.e. loading malicious code into retailers' websites to steal shoppers' credit card details). The Report states that on average over 4800 websites of both big and small businesses are compromised every month, "conservatively yielding tens of millions of dollars to bad actors last year".[17]

In the recent past, the banking and financial sectors world over have witnessed relentless attacks from possible state and non-state actors, organized criminals and hackers. A cyberattack was witnessed by Union Bank of India, where the cyber thieves nearly stole US$ 171 million from the bank. This attack closely resembled the cyber theft of US$ 81 million from the Bangladesh Central Bank's account at the New York Federal Reserve.[18]

According to the Global Economic Crime Survey 2018 conducted by PricewaterhouseCoopers, Cybercrime is the second most reported economic crime globally and financial institutions such as banks serve as preferred target points. Around "49 per cent of organizations globally said they have been a victim of fraud and economic crime – up from 36 per cent".[19] As Cyber criminals find new ways to attack, breach, threat patterns such as phishing and spear-phishing, and social engineering evolve and become more sophisticated. Needless to say, there is a need for robust solutions, enabling organizations to assess their own and their vendors' vulnerabilities in real-time.[20]

III. Threat to Cybersecurity: Four Mechanisms

There are four main dimensions of Cyber vulnerability, or equivalently, four main mechanisms that constitute a threat to Cybersecurity:

- Cybercrimes
- Cyberespionage
- Cyberterrorism, and
- Cyberwarfare

These are briefly discussed below in that order.

A. Cybercrimes

The increasing use of the Cyberspace is today becoming a breeding grounds for Cybercriminals who commit crimes worth billions of dollars. Countries are reporting crimes against critical information infrastructures (CII), which involve enormous damage to life and assets.

The situation in some countries has been worsening with organized crime, such as the mafia networks working through the same space, thereby promoting a cyber underground economy. This money is often used for upgrading malicious codes and software, which could easily be done through cheap and readily available toolkits. Such malicious software has also penetrated into critical infrastructures of countries like Iran. As indicated earlier, Programmable Logic Control (PLC) and Supervisory Control and Data Acquisition (SCADA) systems were targeted by malwares which attacked Iran's nuclear facilities. This was done by a virus called Stuxnet that almost completely destroyed Iran's Natanz nuclear facility by causing the centrifuges to spin out of control.[21]

Moreover, cyber criminals are using innovative techniques through phishing, spams, social networking sites so as to steal confidential and critical information to conduct crimes ranging from financial frauds to espionage. What is most worrisome is the fact that such malicious software seems to be evolved enough now to attack even secure but critical infrastructures. Most importantly, nuclear power has now emerged as a deadly target, which if accessed by ill motives, can cause incalculable destruction.[22]

B. Cyberespionage

Instances of cyberespionage are on the rise with growing reports of data theft, exfiltration (unauthorized copying, transfer and retrieval of data from a computer or server) of intellectual properties worth millions. This phenomenon has been affecting cyber networks of both, governments and private enterprises. Often such incidents aren't reported for the fear of losing the trust of people, in the case of government and clients, in the case of businesses.

Cyberespionage has been penetrating into government systems, hacking into crucial and classified information. What complicates the situation further is that more often than not, the hackers are backed and funded by nation-states themselves. This makes a punitive action exceedingly difficult. All that governments can do is to establish measures and protocols to ensure confidentiality of data. Intelligence agencies typically do not get adequate legal backing in their efforts to secure the relevant networks. Willy-nilly, this ends up favouring cyberespionages and cyber criminals who exploit such loopholes to their advantage.

Here is a recent example of Cyberespionage. A Pakistani App called SmeshApp recently was used to collect intelligence for Pakistan army to attack the Pathankot airbase in India. The Pakistani intelligence first set up "honeytraps" on Facebook by creating fake account users from Airforce, Navy, Border Security Force (BSF), etc. The accounts were used to advertise the SmeshApp on Google Play, which once downloaded, would behave like any other messenger such as WhatsApp. This App created a database of information in the form of photos, contacts, personal data, email, browsing data, etc. Once the Pakistani intelligence agencies extracted the complete data, the mobile phone acted as a microchip tracking the personnel's real-time locations and activities. This helped the intelligence to have a fair knowledge and insights of the Pathankot air base. Shockingly, this App flourished in the user markets with a good rating before it was taken out by Google Play for violating its policies.[23]

C. Cyberterrorism

A grave concern worldwide is the growing menace of Cyberterrorism, wherein terrorists hack into software used for national security, using cheap and readily available software.

Cyberspace has been used as a safe haven for planning terrorist attacks. This space enables terrorists to easily find and attack targets for their perceived furthering of political, social or religious cause. Terrorists use Cyberspace for interaction, communication, infiltration, command and control, propaganda spreading, recruitment and training. The challenge here again is the threat of crimes by non-State actors. The dirty world of terrorism becomes murkier as the Cyberspace is characterized by anonymity, and because such terrorist attacks often lack public acknowledgment.[24]

D. Cyberwarfare

Cyberwarfare covers a wide range of events in the Cyberspace, which forms a part of the so-called Information War (IW). Cyberwarfare basically involves actions by a particular State or its proxies to intrude into another country's computer networks and systems through espionage, damage or disruption. This includes all kinds of media propaganda, perception management and the like.

Countries like the US have already started considering Cyberspace as the fifth domain of warfare, by creating a "force" to prepare for Cyberwarfare and raise its US Cyber Command

(USCYBERCOM). British Government Communication Headquarters has been preparing a cyber force and so has France. Russia too has been actively preparing for a war in the cyber domain. Further, China has been developing appropriate infrastructure since way back in the 1990s, so as to prepare for a defensive cyberwarfare. South Korea too created a Cyberwarfare command in 2009 in response to the Cyber threats by North Korea's Cyberwarfare units. A commonality between all the above major powers is their determination towards building an offensive and defensive force to prepare for a Cyberwar.[25]

IV. Cybersecurity: A Global and Cross-country Perspective

The United Nations first adopted a resolution for "creation of a global culture of Cybersecurity" (UN General Assembly Resolution 57/239) way back on January 31, 2003. This was followed by several amendments.

More recently, while launching its Second Global Cybersecurity Index (GCI), the UN reported:

> "Only about half of all countries have a cybersecurity strategy or are in the process of developing one." The multi-lateral agency urged "more countries to consider national policies to protect against cybercrime." The agency also said that "more efforts are needed in this critical area, particularly since it conveys that governments consider digital risks high priority".[26]

According to the UN, "cybersecurity is an ecosystem where laws, organizations, skills, cooperation and technical implementation need to be in harmony to be most effective".

According to the UN Global Cybersecurity Index (GCI), the top 11 most committed countries include – Singapore, US, Malaysia, Oman, Estonia, Mauritius, Australia, Georgia, France, Canada and Russia, in that order.[27]

The UN report emphasized that there is "space for further improvement in collaboration" at all levels, which advocates for encouraging governments to consider national policies that take into account cybersecurity and encourage private citizens to make smart decisions online.

Against this global backdrop, it would be instructive to examine select country experiences.

1. USA

The American ideals of national security and liberty depend on Cybersecurity to ensure economic growth and unhindered exchange of ideas. Considering the Cyberspace as the fifth domain, the United States established its US Cyber Command in 2011 to defend its interests in the Cyberspace. Today, it has become a global combatant beyond the conventional battle domains of land, sea, air and space.

Reportedly, the US is considering a new act, the National Cybersecurity and Critical Infrastructure Protection Act amending the Homeland Security Act of 2002. This will ensure a partnership between the public and private in handling Cybersecurity.

The US had introduced the use of biometric security measures to access computers, mobile phones, laptops, bank accounts etc. considering passwords to be redundant. Today, many of the mobile phones being manufactured throughout the globe use this pattern of security to prevent cyber threats.

The US now has a cyber resilience policy, which includes Cybersecurity framework especially focusing on critical infrastructures of the nation. It covers Cybersecurity activities such as identifying, protecting, detecting and responding to cyber intrusions. It is also expected to provide feedbacks on the cyber effectiveness, weaknesses and strengths. The US thus, has a Cybersecurity that is self-regulatory and consider the Cyberspace and the threats of utmost importance.

The mandate of the US Cybersecurity policy is to draft strategies for protecting national secrets and developing digital defences, viewing it as a most pressing strategic concern. The Obama administration had asserted that there are cyberattacks by Chinese hackers on Americans and American companies doing business in China, some of them possibly even State-sponsored. Beijing, in its turn, said that it has been a frequent victim of attacks of American origin. Talks between the two countries over cyberattacks and national security leaks have become complicated after revelations that the US National Security Agency spied around the world, even on American allies.

2. China

Today, China houses one of the largest numbers of Internet users and is the second most cyber-targeted nation. Since the times of Jiang Zemin, China has accepted its weaknesses in terms of the Cyberspace and acknowledged the importance of Cybersecurity. China closely watched the US cyber abilities during the Iraq war with the aim to transform itself into a Cyber power.

With the Presidentship of Xi Jinping, the mandate was to draft strategies strong enough to protect national secrets and develop digital defenses. China does undertake surveillance very prominently and cyber manages everything from Internet as a broad arena to communications and connectivity.

3. Israel

Israel is one of the few countries which has heavily invested in Cybersecurity, notwithstanding its small territorial size. Israel believes in rapid innovations in all fields, especially Internet technology. It has created a new Cyber defense infrastructure to bridge the public and private sector in Internet governance.

Israeli Prime Minister Benjamin Netanyahu, in his address to the World Economic Forum 2014, spelt out the details of the nation's impeccable high-tech capabilities and inventions which he stated, is demanded by the entire world. He underscored the importance of securing the global community from cyberattacks.

Israel faces the brunt of its own technological advancements as hacker communities attacking the Israeli networks are often products of their own stealth units. The challenge then arises to secure one's own cyberspace from own technology and limit the hackers from cybersecurity research and solutions. Yet, Israel today is called the nation of Cybersecurity start-ups as it successfully turned the attacks into its source of strength. It faced up to 100,000 cyber-attacks a day with 1-2 serious attacks.

The Israeli government agencies such as the National Cyber Bureau provide tax incentives to these start-ups, which receive investments from throughout the globe. For instance, Israel's national cyber complex in Beersheba received investments from US companies such as Lockheed Martin and RSA Security. These start-ups have established training centres, not just for their own employees, but also for other companies from around the world against network hacking consisting of real-time defense by live instructors.

11

Technology and Surveillance Issues

"Surveillance technologies now available – including monitoring of virtually all digital information – have advanced to the point where much of the essential apparatus of a police state is already in place."

Al Gore[1]

Introduction

Surveillance is essentially a close observation of people. Mass surveillance is not new. It has a somewhat checkered history. A classic example of mass surveillance by a government, over almost all aspects of life, is from the erstwhile East Germany, by Stasi, the State Security Ministry. For nearly four decades, the Stasi, tried to "know all about everyone"[2] through 90,000 full-time employees assisted by 170,000 non-official full-time collaborators, which constituted nearly 2 per cent of the entire East German population. If occasional informers are added, the total number worked out to one person per 6.5 people!

Modern mass surveillance can achieve this level of omnipresence with just a small fraction of the person power. This is because of the incredible sophistication and integration of information and communication technologies (ICTs) in daily lives. The emerging technology has enabled the introduction of extremely powerful modes of mass surveillance. The classical Social Contract theory had assumed that State authorities always and invariably work towards the interest of the people – protecting their lives, liberty and property. This perspective is now being questioned worldwide.

There is no disagreement that a collaborative approach is required among various stakeholders—the government, public institutions as well as the civil society organizations—so as to efficiently draft a surveillance framework as a shared responsibility. Nevertheless, mass surveillance continues unabatedly, world over, with ever increasing technological fineness and intensity.

This chapter deals with the nexus between the emerging New-Age Technology and mass surveillance in the global context. The chapter is organized as follows:

Section I explains the basics of surveillance. Section II places the surveillance effort by governments in the broader context of Social Contract Theories. Against that backdrop, Section III, which is the mainstay of the chapter, elaborates on 'what, how and for whom' of surveillance. The final part, i.e. Section IV presents both sides of the ongoing controversy surrounding surveillance.

I. Surveillance: Basics

As mentioned earlier, surveillance is a close observation of people. This includes keeping track of their habits, behaviour, and actions. Essentially, it involves "monitoring of behaviour, activities, or other changing information for the purpose of influencing, managing, directing or protecting people".[3] Earlier, surveillance was conducted through simple methods such as human intelligence agent and postal interception. Since then, it has advanced into sophisticated digital tools planted in the everyday lives of people. Nowadays, surveillance could mean monitoring from a distance through gadgets, such as Close–Circuit TV (CCTV) or tapping phone calls and text messages. With biometrics and databases, governments can now achieve the scale as desired, without assigning too many people to it. The New-Age Technology has been making it easier than ever before to keep an eye on all pervasive level of surveillance on the people.

Surveillance has been conducted by governments for intelligence gathering, prevention of crime, protection of people or in investigation of crimes. It is used by businesses as well, to gather intelligence say about their competition, and by criminal organizations to plan and commit crimes, such as robbery and kidnapping.

In her book *Superconnected*, Mary Chayko, a Sociologist of Communications Technology, differentiates between two types of surveillance: vertical and horizontal. Vertical surveillance implies that there is a dominant force, such as the government that is attempting to control or regulate the actions of a given strata of society. Such powerful authorities "often justify their incursions as a means to protect society from threats of violence or terrorism. Some individuals question when this becomes an infringement on civil rights".[4]

Horizontal surveillance differs from the vertical one when the tracking by an authoritative source is replaced by an everyday figure, such as a friend, co-worker, or stranger. People leave traces of information in the online world that may reveal their interests, loyalties and desires. While the connectivity "can allow people to become interconnected and develop social connections online, it can also increase potential risk to harm, such as cyberbullying or censoring/stalking by strangers, reducing privacy".[5]

The flip side of mass surveillance is that it is an active infringement on an individual's right to privacy. Not surprisingly, therefore, attempts of mass surveillance are often opposed by civil society organizations and activists. Most functioning democracies have regulations that aim at restricting the use of surveillance technologies, often limiting them to extreme circumstances where national security or public safety is at risk. On the other hand, authoritarian governments rarely, if ever, have checks and balances on mass surveillance, leading to its widespread usage.

In the future, intelligence services across the world might deploy the Internet of Things (IoT) technology for the purpose of identification, monitoring, location tracking and to gain access to private networks. Indeed, in February 2016, the US Intelligence Chief reportedly announced to the American people that "we might use the Internet of Things to spy on you".[6]

In addition, there are other related concepts. For example, Countersurveillance is the practice of escaping surveillance or making it difficult for others to suveil. It has grown dramatically in both scope and complexity with the help of the Internet, increasing spread of electronic security systems and emergence of high-altitude (and possibly armed) Unmanned Aerial Vehicles (UAVs).

Counter-surveillance operations have been used to prevent corporate spying, or tracking criminals belonging to certain crime syndicates. Moreover, it has also been deployed to deter stalking methods used by various entities and organizations.

Inverse surveillance (also called sousveillance) is the practice of reversing surveillance on other individuals or groups (e.g., citizens photographing police). It involves recording by private individuals, rather than government or corporate entities. A well-known example is private recordings of police brutality as in the case of the Rodney King beating.[7] In fact, there is an organization called Copwatch, which attempts to monitor police officers to prevent police brutality.

Evidently, thus, issues relating to the use of emerging New-Age Technology in the process of surveillance is going to occupy a center stage in the times to come.

II. Surveillance and Social Contract Theories

Surveillance, undeniably, violates or infringes upon the basic right to privacy. This raises the fundamental issue of legitimacy of the authority of the State over the individual. Right from the Age of Enlightenment, the relationship between "natural" and "legal" rights of the individual *vis-a-vis* the State has been discussed through successive theories of the 'Origin of the State', especially the theory of "Social Contract."

There are several theories of the "Origin of the State" propounded by Political Scientists. These include, *inter alia,*

i.) The Theory of Divine Right

ii.) The Force Theory

iii.) Social Contract Theory

The Divine Right theory, the oldest among the theories, states that God rules directly or indirectly through a monarch who is not subject to earthly rules.[8] Then came the Force theory which rests on the premise of 'might is right'. According to this theory, force was used by the strong over the weak in order to gain control over them. The stronger group became the master and ruled the weak. The State is born out of force, exists by force and is demolished in the absence of force.[9]

The Lockean Social Contract theory[10] is probably the most widely accepted theory of the "Origin of the State". The Social Contract theory argues that the modern State owes its genesis to a contract between all men (and women) with each other.

According to John Locke, before the State came into being there existed a "State of Nature" where everybody enjoyed a peaceful natural life. The laws of nature governed humans. However, since there was no common-superior in power, every individual came-up with his or her own interpretation of the natural law. This fueled fear and danger in society. It was to resolve this problem that people entered into two contracts.

First, the Social Contract which established a peaceful community with the right to property. Second, the Political Contract between the people and the government which established a Constitutional Monarchy. Under this system, the government is only a trustee and the King could be dethroned if he failed to provide safety and security to the people.

The natural right to life, liberty and property rested with the people themselves.

Right to privacy is a fundamental right encompassed within the right to life[11] which was not conceded in the Lockean contract. The *sui-generis* problem of Surveillance State could not possibly be thought of in the age of Locke. The change of circumstances, owing to technological advancement, has made the Lockean Social Contract invalid, i.e. unable to address the wrongdoings, if any, perpetrated at the hands of the State.

Two solutions to this problem have been proposed:

First, one is by Simon Chesterman who maintains that the Lockean contract cannot solve modern day problems of the Surveillance State and argues for a new Social Contract. In this new Social Contract, individuals give power to the State to access their personal information in exchange for security and public utility services in the modern world. The cardinal difference between the two Social Contracts is the right to privacy being yielded in exchange for a safe living environment.[12]

Transnational threats and the need for intelligence services in criminal proceedings seem to underscore the requirement of a modern Social Contract consisting of consented access to private information by the State. Unlike Locke, Chesterman's Social Contract has an opt-out clause. Herein, if a person decides not to have an active digital life and chooses not to avail public utility services then the person will not have a digital footprint and accordingly could not be surveilled. The essence of the problem here is that people often see privacy and security as a mutually exclusive choice. The imperative need to strike the right balance between the two cannot be emphasized more.

Second, Tim Berner Lee, founder of the World Wide Web, has offered to restore the Lockean contract with a supplementary contract on the specific issue at hand. He proposed a Social

Contract exclusively for the World Wide Web with duties for each of the stakeholders to uphold the principles of individual privacy and data protection.[13] This contract entrusts responsibilities to governments, companies and the citizens for collectively protecting the rights of all and promoting humane use of technology among other goals.

III. Surveillance: What, How and For Whom

With the revolutionary advancement in technology, mass surveillance has become easier and low-cost than ever before. Various means of surveillance that have sprung up under the emerging New-Age Technology may be described as follows:

A. Telephone Surveillance

Tapping of telephones lines has been widespread since the advent of the technology. For example, in the USA, the relevant Act requires that all telephonic communications must be available for real-time wiretapping by Federal law enforcement and intelligence agencies. Two major telecommunications companies in the US reportedly, have contracts with the Federal Bureau of Investigation (FBI) that require them to keep their phone call records easily searchable and accessible for Federal agencies. Interestingly, human agents are no longer required to monitor most calls. The speech-to-text softwares can develop text which is readable by machine and is intercepted from audio, which then is processed by automated call-analysis programs.

Law enforcement and intelligence services in the US and the UK have access to technology that can activate microphones in cellphones through remote based locations that can help them track conversations taking place near the person who holds the mobile phone. This is surveillance through remote eavesdropping.[14]

Cell-phones have been used to collect data. Geographical location of mobile phones can be determined even when it is not in use. In response to privacy concerns emanating from the infamous Snowdenleaks, newer models of Apple's iPhone have been designed to disrupt investigative wiretapping efforts. The Snowden leaks demonstrated that US agencies have the ability to access phones anywhere in the world. This led to privacy concerns in countries with growing markets making it imperative for companies, such as Apple, to address those concerns to secure their position in the global market. New iPhone models have thus locked out the National Security Agency (NSA) of the US.[15]

Reportedly, the NSA has been taking advantage of the ambiguity in the relevant law by collecting metadata on "at least hundreds of millions" of "incidental" targets from around the world.[16] The NSA has deployed an analytic tool known as CO-TRAVELER that helps track people, whose movements intersect for finding hidden connections with persons of interest, towards law enforcement and crime solving purposes.[17]

B. Camera Surveillance

Surveillance cameras have been installed by the millions in many countries, and are now monitored by automated computer programs instead of humans. These are video cameras that observe an area and are often connected to a recording device or network that is watched by a law enforcement officer.

Cameras and recording equipment used to be relatively expensive earlier and required human personnel to monitor the footage. With the emerging technology, analysis of footage has been made easier by automated software that organizes digital video footage into a searchable database, and by video analysis software (such as VIRAT and HumanID).

As of 2016, there were about 350 million surveillance cameras worldwide, about 65 per cent of which were installed in Asia. In 2018, China was reported to have a huge surveillance network of over 170 million CCTV cameras with 400 million new cameras expected to be installed in the next three years, many of which use facial recognition technology.[18]

Notably, these surveillance cameras along with advanced video analytics and facial recognition software can identify and track individuals wherever they go. The cameras are reportedly connected to a centralized database and monitoring station, which contains a picture of every person's face in China – over 1.3 billion people.[19]

C. Computer Surveillance

Computer surveillance typically involves monitoring of data and traffic on the Internet. In the US for example, under the relevant Act, all broadband Internet traffic (emails, web traffic, instant messaging, etc.) are required to be made available for real-time monitoring by the Federal law enforcement agencies.

There is exceedingly large and growing data on the Internet – far too much for human investigators to manually search through. Thanks to the emerging technology, automated Internet surveillance computers have been devised now, which go through the vast amount of intercepted online traffic to identify and report any suspicious happenings to human investigators. This process is based on identifying certain "trigger" words or phrases and operative systems to intercept and analyze the data collected. This is then used to extract information, which could be useful to law enforcement and intelligence agencies.

Computers are a surveillance target simply because personal data is stored therein. If someone is able to install a software, such as the FBI's Magic Lantern on a computer system, they can easily gain unauthorized access to this data. Such software could be installed physically or even remotely.[20] Interestingly, the NSA runs a program called PRISM, which is a data mining system that gives the US government direct access to information from technology companies. By accessing this information, the

US government is able to obtain search history, emails, stored information, live chats, file transfers, and more.[21] Not surprisingly, the program triggered major controversies regarding the trade–off between surveillance and privacy.

D. Social Network Analysis

Social network analysis i.e. analysis of social networking sites such as Facebook, enables governments to collect detailed information about peoples' friends, family and other connections. Since much of this information is voluntarily and unwittingly made public by the users themselves, it is nothing short of being "open-source intelligence".

Many US government agencies have been investing heavily in research involving social network analysis. They seem to believe that the biggest threat to the US power comes from decentralized, leaderless, geographically dispersed groups of terrorists, subversives, extremists, and dissidents. Such threats could be countered by finding important nodes in the network, and removing them.[22]

E. Biometric Surveillance

Biometric surveillance measures and analyzes human physical and/or behavioral characteristics for authentication, identification, or screening purposes. Physical characteristics include fingerprints, DNA, and facial patterns, whereas behavioural characteristics include voice or gait (a person's manner of walking).

Facial recognition is based on the unique configuration of a person's facial features to accurately identify him or her, usually from surveillance video. New technologies are now capable of identifying a person by their facial features from a distance of upto 500 ft.

Yet another tool here is DNA profiling, which basically examines some of the major markers in the body's DNA to produce a match. The FBI is reportedly in a process of building a new biometric database, which will store DNA, facial recognition data, iris/retina (eye) data, fingerprints, palm prints, and other biometric data of people living in the United States.

F. Aerial Surveillance

Aerial surveillance basically involves collecting surveillance data, usually visual imagery or video, from an airborne vehicle—such as an unmanned aerial vehicle (UAV), helicopter, or spy plane. Military surveillance aircrafts use a variety of sensors (e.g. radar) to monitor the battlefield.

Digital imaging technology, miniaturized computers, and numerous other technological advances over the past decade have contributed to rapid advances in aerial surveillance hardware such as micro-aerial vehicles, and high-resolution imagery capable of identifying objects at extremely long distances.

IV. Surveillance Controversy

A. The Case for Surveillance

Those who defend surveillance systems argue that these tools can help protect the society from terrorists and criminals. According to them surveillance can contain crimes through at least three channels: by deterrence, by detection, and by reconstructions.

First, surveillance can be a deterrent by enhancing the probability of terrorists and criminals being caught, and by revealing their *modus operandi*. Secondly, surveillance can be used to fight criminals or terrorists by detection i.e. by linking the information stream obtained through surveillance to a recognition system. For example, a CCTV feed run through a facial recognition system may be able to auto-recognize and detect the criminals or terrorists and direct the relevant investigative authorities to their location. Thirdly, surveillance can be used for reconstruction of a criminal or anti-national activity. Surveillance can provide human operatives a tactical advantage through improved awareness of the situation or through the use of automated processes, such as video analytics. In the process, surveillance can help reconstruct an incident and prove the offence by making available the relevant footage for forensics experts.

Some surveillance systems also have other uses. For instance, it can help locate and retrieve runaway children, abducted or missing adults as well as missing mentally challenged people.

Some supporters of surveillance believe that besides identifying and capturing or retrieving individuals as indicated above, surveillance also tends to create in many people a feeling of always being watched. As a result, they may become self-policing. This allows the State to control the entire populace without having to resort to large physical force, which is usually expensive.[23]

According to Foucault, the concept of 'panopticism' is a means of indirect control over a large populace through the uncertainty of surveillance. He analyzed the architecture of the prison panopticon – a circular prison design in which prison guards are located within the center of the facility, monitoring the prisoners who are invisible to one another, and realized that its success was not just in its ability to monitor but also its ability to not monitor without anyone knowing.

The metaphor of 'panopticon' has been used to illustrate the influence of surveillance on an individual, in addition to the society as a whole. Although the inmates are not consistently observed, they are nevertheless obliged to abide by the rules due to the perception that they are under surveillance.[24] "Feeling of being always watched" engendered by surveillance works in the same way.

B. The Case Against Surveillance

With the arrival of high-tech devices such as high speed surveillance computers and biometrics software, governments now have unprecedented ability to monitor the activities of their subjects.[25] Many civil rights and privacy groups in the US have expressed serious concern that by allowing a steady increase in government surveillance of citizens "we will end up in a mass surveillance society, with extremely limited, or non-existent political and/or personal freedoms".[26] Fears such as this have led to numerous lawsuits.

Most opponents of surveillance do not wish to sacrifice their privacy. For instance, a person who is part of a political group which opposes the policies of the national government, might not want the government to know his or her name or what their like-minded friends have been reading, so that the government cannot easily subvert their organization, arrest or kill them.

Several critics have pointed out that surveillance for the purpose of national security can often be misused for an illegal and unauthorized interception. This is possible only because of the lack of accountability when it comes to surveillance and intelligence agencies.

There is evidence to suggest that a heavily bureaucratized and minimally accountable regime does not necessarily contribute to enhancing security while significantly costing the people's privacy. For instance, while looking at the US Mass Surveillance Program of the National Security Agency (NSA), an American court found that "not even a single successful pre-emption was based on material collected from the NSA's surveillance regime out of more than 50 instances where terrorist attacks were prevented".[27]

In addition, it has been argued that surveillance engenders so much racializing quality that it operates almost as "racializing surveillance." Browne uses racializing surveillance to refer to "moments when enactments of surveillance are used to reify boundaries, borders, and bodies along racial lines and where the outcome is discriminatory treatment of those who are negatively racialized by such surveillance."[28]

It goes without saying that there is an imperative need for a legislation with provisions on how to deal with surveillance, while striking a balance between privacy and national security. Mass surveillance seriously undermines the trust between the citizens, government and judiciary, and creates a negative perception in the minds of people. No amount of unaccountable surveillance, however sophisticated it may be, can possibly justify the feeling of insecurity in the people's mind. Targeted limited surveillance with well defined (and possibly transparent) accountability seems to be the lesser of the evil.

12

Technology: Weaponization and Threat to National Security and Global Peace

"Moore's Law of Mad Science: Every 18 months, the minimum IQ necessary to destroy the world drops by one point."

Eliezer Yudkowsky[1]

Introduction

Walter Lippman, an American Writer and Political Commentator who reportedly coined the phrase "Cold War" once observed:

"A nation has security when it does not have to sacrifice its legitimate interests to avoid war, and is able, if challenged, to maintain them by war".[2]

Indeed, "National Security" is the primary concern of any nation-state around the world. It involves the well-being of every individual citizen in the country. Extending this notion further, 'Global Peace' may be seen as involving transnational security concerns and therefore, it covers the well-being of the entire humanity.

This chapter is focused on global policy issues in national security and global peace. It has been organized as follows:

Section I explains the meaning of national security and elaborates on its various dimensions. Section II is devoted to the ongoing massive weaponization, enabled across the world, by the emerging post-Digital Revolution Technology. Against that backdrop, Section III discusses a range of Public Policy issues brought up by weaponization of the New-Age Technologies, especially those relating to the threats to national security and global peace.

I. National Security: Meaning and Dimensions

National Security was originally conceived as protection against external military attacks. It essentially meant freedom from foreigners dictating the governance of the relevant nation-state. With changing times, this notion of national security has changed considerably. Indeed, the scope of national security has continually been widening.

In the early 1980s, Harold Brown, the US Secretary of Defense, defined national security as: "The ability to preserve the nation's physical integrity and territory; to maintain its economic relations with the rest of the world on reasonable terms; to preserve its nature, institutions, and governance from disruption from outside; and to control its borders".[3]

More recently, national security is seen as capacity of a nation-state to address and overcome the potential multi-dimensional threats to the well-being of its citizens and its geo-political survival.

There are several dimensions to national security. These dimensions may be grouped under two inter-related categories:

a.) Geo-political Security

b.) Socio-economic Security

For a nation-state, geo-political security may be threatened by a host of factors which, *inter alia*, include:

- Military attacks
- Terrorist attacks
- Cyberattacks
- Natural disasters (e.g. flood, earthquakes, etc.)
- Climate change

On the other hand, socio-economic security obviously comprises social as well as economic security. Social security could be threatened by (i) deterioration or absence of the so-called "law and order" and (ii) marginalization of some communities (e.g. social boycotts and atrocities on some strata of the society).

Economic security, in a way, subsumes the basic social security (i.e., access to basic human needs of education, health, shelter and social protection) besides work-related security.[4] Work-related security, inter alia, consists of:

1. Income Security – i.e., adequate actual, perceived and expected income, either earned or in the form of social security and other benefits.
2. Labor Market Security – i.e. availability of ample opportunities for adequate income-earning.
3. Employment Security – i.e. protection against loss of income-earning work.
4. Job Security – i.e. ability to pursue a line of work in conformity with one's interest, training and skills.[5]

In sum, national security may be seen as a combination of several factors, including but not limited to, human resources, natural endowment, political resilience, economic structure and strength, technological competence and, of course, the military might.

While national security is widely regarded as a matter of the Rule of Law and good governance, the concept is not unlimited. In fact, way back in 1968, in the case: *United States v. Robel*, the US Supreme Court had ruled that the US government cannot deprive the people of their Constitutional rights, even in the interests of national security.

Indeed, according to the historic judgement:

> *This concept of national defense cannot be deemed an end in itself, justifying any exercise of legislative power designed to promote such a goal. It would be ironic if, in the name of national defense, we would sanction the subversion of one of those liberties which makes the defense of the Nation worthwhile.*[6]

Later day skeptics have gone even further. For example, Ralph Nader, an American politician and author regards national security as the "fig leaf against freedom of information".[7]

Be that as it may, there is no denying that there is too much animosity and hatred in the world. This makes building a strong national security framework a top priority for every nation-state.

The potential impact of the New-Age Technology on socio-economic security was discussed in chapter 5. In this chapter, the attention has been focused on threats to geo-political security and global peace, arising from the ongoing weaponization engendered by the emerging technologies, which is taken up next.

II. Weaponization of New-Age Technology

Much of the New Age Technology i.e. the post-Digital Fusion Technology, has been emerging out of the growing research in military and defense establishments of the leading nation-states, especially the US, Russia and China. Not surprisingly, therefore, there is a massive ongoing weaponization of the emerging technology.

The focus here is on "what and how" of the nexus between the New Age Technology and weaponization that is currently underway. The coverage here includes: Lethal Autonomous Weapon Systems, Additive Manufacturing (especially, 3D Printing Technology and how it is revolutionizing the creation of self-made weapons), Autonomous Robots being used as soldiers, and new weapons based on the technology of Augmented Reality.

A. Lethal Automated Weapon Systems

The International Committee of the Red Cross defines "autonomous weapons" as an umbrella term encompassing any type of weapon with "autonomy" in its "critical functions", meaning "a weapon that can select (i.e. search for, or detect, identify, track) and attack (i.e. intercept, use force against, neutralize, damage or destroy) targets without human intervention".[8]

Consider, for example, the weapon systems used by Navy. The application of Artificial Intelligence (AI) in partially automated Naval Combat Systems increase human-machine collaboration as it augments the decision-making process of the officers on board, "by intelligently processing multiple sources of information,.... with existing sensors to assess and confirm potential threats".[9] AI helps to "improve operational capability in its fighting units and enables rapid decision-making in complex, fast-moving war fighting scenarios".[10]

AI has also led to the use of fully automated unmanned vessels which are remotely operated. A remotely placed human feeds in the necessary command with appropriate coding, while supervising the movement of the vessels using satellite links.[11] AI helps autonomous vehicles to execute a task without human input.

Submarines, frigates, aircraft carriers and battleships, can all be equipped individually with the Command, Control, Communications, Computers, Intelligence, Surveillance and Reconnaissance – the so called C4ISR – oriented system so as to substantially enhance their specific functions. Through neural networks and deep learning algorithms, the AI-based intelligent systems help recognize behavioural patterns, carry-out appropriate simulations and thereby, enable end-users to greatly improve detecting and tracking of combatants.[12]

Among private players, Rolls Royce has announced its intention to develop completely autonomous ships, devoid of any human crew. The manufacturer has "incorporated the latest in navigation technology, combining an array of sensors with an AI-powered computer".[13] The company, reportedly, will use cloud learning engine to train its AI-based object classification system.[14] This software will primarily be used for detection, tracking and identification of potential dangers.

When it comes to aerospace, an up and coming concept is the Swarm Drones Technology, wherein a large number of drones are deployed simultaneously and collectively for a task. Being AI-powered, they are capable of working fully autonomously. Incidentally, the Swarm Drone Technology is being extensively used by the US Navy for diverse tasks, ranging from laying and destroying mines to protecting warships and submarines.[15] In fact, the US has an entire programme dedicated to this, titled "Low Cost Unmanned Aerial Vehicle Swarming Technology" (LOCUST).[16] In response to the US initiative of Swarm Drones technology, reportedly, Russia has been developing an anti-swarm technology called "Repellent".[17] In this technology, drones have full autonomy to execute functions under human supervision, but without continuous remote control.

Artificial Intelligence (AI) also helps in field operations and in combat situations where patience, skill, and navigational expertise is needed.

B. Additive Manufacturing (3D Printing)

In 2013, a US citizen named Cody Wilson, who described himself as a crypto-anarchist and also a gun-rights activist, posted online free blueprints for a one-shot plastic pistol that could be made by a 3D-printer with the help of Additive Manufacturing (AM). The pistol was called the Liberator. The blueprints were downloaded over 100,000 times before the US State Department moved in. Department ordered Cody Wilson to take down the files that were uploaded. The US Department of State claimed that "Wilson violated arms export statutes, even though Wilson says there is no law that says publishing gun design online is criminal".[18] The crypto-anarchist, in turn, termed these orders by Department of State as a restriction on his first amendment rights i.e. freedom of speech. He argued that the blueprints fall within the ambit of 'speech'.[19] He sued the Government and won!

The threat of self-made weapons is not new. Until recently, however, people didn't have the wherewithal and skills to make them as good as the weapons which are commercially available.[20] Shockingly, now it is all a matter of mere printing. Additive Manufacturing will make it easier for individuals to manufacture weapons that could seriously jeopardize the law and order situation in any country. More recent articles describe the complete fabrication of a .22 caliber pistol (excepting the precision metal rifle bore) via 3D printing.[21] The final system was able to shoot 200 rounds with no evidence of wear or tear. Evidently, more complex, perhaps even enhanced weapons will be made eventually by individuals.

There is no doubt that AM has security risk in terms of maintaining the law and order situation. However, it is also deemed to be the future of defense sector. Illustratively, Lockheed Martin has successfully flight tested a Trident II D5 Intercontinental Ballistic Missile (ICBM) with a built-in 3D-printed component (a connector backshell).[22] According to the company, the time required to manufacture military satellites could be reduced by 40 per cent over the next three to five years, thanks to advancements in metallic AM processes. At another plain, China 3D printed titanium components for its carrier-based multipurpose fighter aircraft called the Shenyang J-15 (also known as Flying Shark).[23]

C. Autonomous Killer Robots

Autonomous Killer Robots are considered to be the next generation of weapons enabled by the emerging technology revolution.[24] Israel and South Korea lead the world by having already deployed armed Robots.[25]

It is sometimes argued that these weapons would not be able to operate within the current laws of war, and liability on wrongful death will be uncertain. On the contrary, it is also contended that these weapons will actually reduce human casualties on account of their precision. At the same time it was argued in a Conference of United Nations Institute for Disarmament Research that "the asymmetric use of lethal Autonomous Robots on a battlefield could mean that rather than fighting robots, attacks may be directed at civilian populations."[26]

Interestingly, the UN Special Rapporteur on Extrajudicial Execution in Pakistan has called for a pre-emptive ban on the development of autonomous weapons systems.[27] Such technology, it is argued, "might allow leaders to engage in conflicts without risking soldiers' lives. This could mean that military options might be chosen over a policy of dialogue and avoidance of conflict".[28]

D. Augmented Reality Systems

The US Military's DARPA (Defense Advanced Research Projects Agency) unit has been developing a very special Augmented Reality System for the battlefield. This system allows commanders to send maps and other relevant information directly to the soldiers' field of vision. This gadget is attached to a military helmet, and it is also possible to integrate it with the weapons control system. Here, "instead of looking down at a 2D map or chest-worn computer, the soldier sees virtual icons (such as navigation waypoints, friendly forces, and aircraft) overlaid on their real-world view".[29]

III. Weaponization of Technology and Threats to National Security and Peace: A Global Perspective

As noted before, there is a massive ongoing weaponization enabled by the New-Age Technology which has been posing serious threats to national security and global peace.

It is increasingly being recognized globally that nation-states cannot address their own national security issues in isolation. Individual national security strategies, have to per force, take on board, security of the region as well as the international context.

Under the umbrella of United Nations (UN), the First Committee of UN General Assembly, called the Disarmament and International Security Committee (DISEC), deals with disarmament and international security matters.

The Committee has two main bodies: (a) the UN Disarmament Commission (UNDC) and (b) the Conference on Disarmament (CD), both reporting to the UN General Assembly (through the Committee). The Conference on Disarmament, though technically not a part of the UN, reports to the UN General Assembly.

The Conference on Disarmament which meets in Geneva at least thrice a year, is focused, [30] *inter alia*, on the following:

- Cessation of the nuclear arms race and nuclear disarmament
- Prevention of nuclear war
- Prevention of an arms race in outer space, and
- New types of weapons of mass destruction and new systems of such weapons

There are two facets of weaponization arising from the emerging technologies. First, when it is used against the nation-state for disrupting the law and order situation, and second, when it is used by the nation-state itself in the interest of national security. It is important to distinguish between the two. The former does not know any boundaries whereas the latter has to work within the confines of the Constitution of the nation-state concerned.

As far as the "Autonomous Weapons" technology is concerned, it is required to comply with various international laws, especially Article 36 (Additional Protocol I) of the Geneva Convention, which necessitates nations to do a legal review of new weapons to ensure that they are in full compliance with international laws. The Laws of War, encoded in Geneva Convention and Rules of Engagement, prescribe what is, and what is not acceptable in the battlefield.

United Nations has held several rounds of consultations under the United Nations Convention on Certain Conventional Weapons (UNCCCW), but could not arrive at a conclusion about regulating these weapon systems. Consequently, a group of NGOs have initiated a campaign – the International Campaign to Stop Killer Robots, demanding a pre-emptive blanket ban on the development and deployment of Autonomous Weapons until the concept of "meaningful human control is well documented and legalized by (nation) states".[31]

For all Artificial Intelligence (AI)–driven weapons, the relevant technology must be developed keeping in view various standards of International Humanitarian Law including and prioritizing the "principle of distinction", i.e. ability to discriminate between combatants and non-combatants, so as to minimize civilian damage in an automated attack. AI can be coupled with already available technologies and can provide greater target discrimination to AI-driven weapons.[32]

Some analysts suggest that Robots should possess "moral intelligence". A moral agent is one that adheres to a system of ethics which it complies with, while taking a particular action or refraining from it. Such a policy will compel roboticists and software developers to build ethical architecture for unmanned systems in the future. We should strive to create such an ethical architecture for unmanned systems, where "moral intelligence" is derived from International Conventions.[33] As the moral basis, autonomous robots should subscribe to the principle of Proportionality. This principle argues for the necessity of intentionally reducing civilian casualties as far as possible.[34]

National security implications of the weapons produced using the 3D Printing Technology are obviously enormous. These developments raise a host of concerns for law enforcement and anti-terrorism organizations as 3D printers may enable production of firearm parts which normally require a license to purchase. An even more worrisome dimension involves the Additive Manufacturing Technology being used to render detection of nuclear proliferation more difficult.[35]

It is abundantly clear that countries have been trying to use AM for military purposes. Importantly, however, just one leak or hack of the blueprint could cause a worldwide devastation. If it gets in unwanted hands, it would in all probability, be used to promote radicalism and terrorist activities, with disastrous consequences. Clearly, therefore, addressing criminal and legal concerns would require active cooperation across multiple agencies involved in national security.

Tomorrow's defense and national security workforce will require a new set of advanced skill-sets that will have to be built into their professional training and career development. In this regard, an exciting development is the use of virtual and Augmented Reality (AR) Systems. These tools reduce costs and mitigate risks associated with live action training, and they can be crafted to simulate and visualize events in space and cyberspace to enhance workforce proficiency. In this way, by combining talent and technology, defense and national security agencies could be ready for the challenges that lie ahead.[36]

13

Technology: Regulation and Strategic Development

Introduction

The jury is still out on whether the emerging New-Age Technology underlying the Industrial Revolution 4.0, has a cost-benefit ratio that is favorable for the future of the humankind.

Elon Musk, Founder CEO of Space X and a major player in emergence of the New-Age technology inter alia said:[1]

"Robots will do everything better than us."

"The global arms race for Artificial Intelligence (AI) will cause World War III."

"AI is a greater risk than North Korea."

"Artificial Intelligence is a fundamental risk to the existence of human civilization."

At the other extreme, Mark Zuckerberg, CEO of Facebook, says:

"I think you can build things and the world gets better. But with AI especially, I am really optimistic. And I think people who are naysayers and try to drum up these doomsday scenarios... It is really negative and in some way I actually think it is pretty irresponsible".[2]

The AI Head of Google, John Giannandrea, objected *"to the hype and the sort of sound bites that some people have been making, I am definitely not worried about the AI apocalypse".*[3]

Whatever be the final verdict on the cost-benefit analysis of the emerging New-Age Technology, it really does not matter. The transformative revolutionary technology is bound to co-exist with human lives, no matter what. A more relevant and practical question that needs to be addressed urgently, is: therefore what is the optimal strategy for regulation and development of the unfolding New-Age Technology.

This is precisely the focus of this chapter. In this regard, while there are several important issues that merit discussion, the attention is focused here on:

a.) Regulation and Development of AI (Section I)
b.) Data Localization Issues (Section II)
c.) International Cooperation Issues on Data Sharing (Section III)
d.) International Cooperation Issues on Privacy (Section IV)
e.) Banking-Finance 4.0: Regulatory Issues (Section V)

I. Strategies for Regulation and Development

Most advances in the New-Age Technology have been occurring at breakneck speed. This makes it hard for regulation to keep up. This is certainly true in the case of Artificial Intelligence (AI). A stream of Industry-specific advances in AI have been springing up without any fuss, while its regulation may take months (if not years) before becoming operational. In the emerging post-AI world, where AI drives most aspects of human life, as Elon Musk would probably put it, "an existential" question for human beings that needs to be addressed on priority relates to the optimal strategy for regulation and development of the AI. Needless to say, the time is indeed most propitious for a serious thinking about regulating the development of AI to the extent possible, and shaping the future of humankind.

A. AI Strategies: A Cross-country Perspective

All the leading countries in the world have come up with their own country-specific strategies for regulating the development of Artificial Intelligence (AI).

(i) United States

The United States, the undisputed leader of the free world, surprisingly did not have a well-designed AI strategy, until as late as February 2019. Actually, a beginning was made in the last year of the Obama Administration. In 2016, the then President Obama formed a Sub-Committee on Artificial Intelligence and Machine Learning, which led to the "creation of three globally influential reports: "Preparing for the Future of AI", "The National AI Research and Development Strategic Plan" and "AI, Automation, and the Economy".[4]

Not much happened thereafter for a while. In fact, pressure was building up on the Trump Administration for formulation and announcement of the US AI strategy.[5] Almost three years after the Obama initiative, the US President Donald Trump, finally issued an Executive Order on February 11, 2019 launching the US Initiative on AI, summarized as follow:

The American AI Initiative is guided by five principles, which include the following:

1. Driving technological breakthroughs,
2. Driving the development of appropriate technical standards,
3. Training workers with the skills to develop and apply AI technologies,
4. Protecting American values including civil liberties and privacy and fostering public trust and confidence in AI technologies,
5. Protecting US technological advantage in AI, while promoting an international environment that supports innovation.[6]

It is abundantly clear now that the US aims at maintaining its leadership in research and development of Artificial Intelligence.

(ii) Russia

Russia has consistently been serious about AI. In the UN forums, for example, Russia has been dismissive of ongoing deliberations to prohibit lethal autonomous weapons and has openly advocated military use of AI.[7]

In fact, Russian President Vladimir Putin famously said in September 2017:

> "Artificial intelligence is the future not only of Russia but of all of mankind. There are huge opportunities, but also threats that are difficult to foresee today. Whoever becomes the leader in this sphere will become the ruler of the world".[8]

Putin added it would be better to prevent a monopoly in the field, and said that *"if Russia becomes the leader in the development of AI, we will share our technology with the rest of the world, like we are doing now with atomic and nuclear technology."*[9]

Of course, one doesn't know whether this is a promise or a threat.

Just a month before US President Trump announced the American AI strategy, in January 2019, President Putin approved a list of instructions (for the Agency for Strategic Initiatives) which, *inter alia*, included the instruction to create a national AI strategy for Russia.

The Russian AI strategy is being formulated on the basis of a 10-Point Plan that was approved earlier in 2018. The 10-Point Plan "includes forming an AI and big data consortium across educational and industrial organizations; developing a fund to help provide expertise on automated systems; increasing state support of AI training and education; building an AI lab at a top technology university; establishing a national AI R&D center; monitoring technical and social trends in AI; organizing AI war games; discussing AI proposals at domestic military forums; and holding an annual AI conference".[10]

It is pertinent to note that none of those points convey anything that would even remotely suggest government-controlled or regulated development of AI. It is aimed only at strengthening the AI research. Presumably, therefore, the Russian approach is no different from the American mission of staying ahead in the ongoing AI race.

(iii) China

China also does not seem to convey any intentions of regulating AI. Interestingly, even somewhat uncharacteristically, China has laid down a comprehensive blueprint of how it plans to win the AI race in its "Next Generation Artificial Intelligence Development Plan" (July 2017). This policy outlines China's strategy to build a domestic AI industry worth nearly US$ 150 billion in next few years. China plans to bring its AI industry "in line" with its competitors by 2020, and thereafter plans to lead the world in certain AI sectors by 2025 and finally aims at becoming the leader in AI research by 2030. This document "officially marked the development of the AI sector as a national priority in China and it has been included in President Xi Jinping's grand vision for China".[11]

With the US, Russia and China—the three big powers—aiming for the top slot in the development of AI, a sort of new Cold War is seemingly emerging.

In this tri-polar Cold War (or really, bi-polar Cold War with Russia and China on one side, and the US on the other), there is a view that China has already taken a lead in AI research.

Symptomatically, after analyzing more than 2 million research papers on AI published by end-2018, Allen Institute for AI found that China is "poised to overtake the US in the most-cited 50 per cent of papers this year, in the most cited 10 per cent of papers next year, and in the 1 per cent of most-cited papers by 2025".[12] It turns out that the US share of the most cited 10 per cent of research papers declined from 47 per cent in 1982 to 29 per cent in 2018. During this period, China's share rose sharply reaching 26.5 per cent.[13]

On the other hand, there is a view that the US is "not doing enough to stay ahead in AI." President Trump's Executive Order of February 2019 is deemed to be "vague, with few concrete goals, and included zero new funding for research. Other countries' national AI strategies have included government investment ranging from US$ 20 million (in Australia and Denmark) to nearly US$ 2 billion (in South Korea)".[14]

President Trump's initiative, arguably, failed to address the biggest challenge in the AI field i.e. "attracting global talent. The number of top-flight AI researchers is limited and the current US approach to immigration is the opposite of what is needed".[15]

To put it differently, the AI race to the top has implications not only for global peace, but also for all the functioning democracies in the world.

B. Regulation of AI

As far as regulation of AI is concerned, the policy-makers around the world would have to be proactive rather than reactive. Generally policies are often framed as a reaction to events. For instance, the Patriot Act by the US Administration under President Bush was a response to the 9/11 attacks[16] that had shaken the US and the world. Two observations are noteworthy here. First, it took a month after the attack for the Patriot Act to be put together. The formulation of regulation in all democratic countries, is a much slower process and can take several months of deliberation before becoming a law. This kind of a timeline would simply not be desirable in the world where large-scale changes can take effect in just a few days and could have the potential to bring about a major disruption in the industry. In other words, there is a need to speed up regulation to cope with the pace of AI development. Secondly, regulation typically tends to be everlasting in a world of continuous and accelerated change. So if an Act is passed, it is likely to be law for several years before it becomes virtually obsolete and is either revised or repealed.

In other words, not only does the regulation tends to be reactive in practice, the duration of reaction time can also be too long. This is a major problem going forward, especially in the AI realm, because chances are that by the time a reactive law is passed, it may already be too late.[17]

There are two ways in which tech policy, and in particular AI regulation, can be made proactive:

a.) Using Mandatory Sunset Clauses: Having an expiry date on laws would ensure that there is a periodic steady stream of revision of laws in keeping with the pace of AI development.

b.) Maintaining Level Playing Field: Another way in which AI advances can be kept in check is through maintaining standards in various sectors. This is so because if left unchecked, it could spell problems for competition. For instance, "if your competitor is rushing to build AI and you don't, it will crush you. It is the job of the regulator to ensure that advances in AI don't tend to push sectors towards a monopoly." [18]

C. Legal Issues

There are a number of legal issues plaguing the use of AI. Probably the most important question in this regard is who would be liable for the damage to life or property caused on account of a failure or malfunction of an AI?

AI systems are simply machines, which can malfunction and cause error at times. Though there are continuous improvements taking place in the areas of cloud-based infrastructure, there is a need to debate, discuss and address the issue of liability.

A prime example could be that of Tesla, the car manufacturer who brought 'autopilot' feature in their self-driving cars based on AI. Now, suppose that a person driving the car activates the autopilot feature and during such time, an accident occurs, two questions arise: (a) Is the software developer (person/ organization who installed/designed this feature into a vehicle) to be held liable? Or, (b) can AI itself be held liable?

To answer the first question, the concept of 'negligence' has to be looked into. Negligence means failure to take proper care of something one ought to have taken care of. In order to prove 'negligence', four elements must be established: 1) Defendant had a duty of care, 2) Defendant breached the duty, 3) Breach caused an injury to the plaintiff, and 4) the harm caused to the plaintiff was such that it could have been anticipated at the time of making of the product.[19] However, AI systems are designed to keep learning from the datasets available to them, the probability of foreseeing the harm would be very small. Therefore, holding a developer accountable for the actions of AI would not be appropriate.

For the question of liability of AI, there is seemingly no settled position in several countries with regard to its 'personhood' as yet. The US position is that Robots cannot be sued even though they can cause damage to the life or property.[20] There are others who think that AI entities should not be considered as subjects of law.[21]

In short, the liability with regard to the actions of AI remains uncertain. Unless and until, the nation-state by way of law, passes an Act clarifying the same, it would be difficult to attribute the liability to AI or the developer.

II. Data Localization Issues

Data Localization is defined as the act of storing data on any device that is physically present within the borders of a specific country from where it was generated.[22]

There are many ways to locally store data, some more stringent than others. Data Localization can be divided in four main categories:

(a) Complete Restriction on Data Export

This category allows no data or its copies to leave the country generating it. It compels foreign companies to create or rely on servers located within the country and develop infrastructure to maintain the data. Not surprisingly, China and Russia deploy variants that fall under this category.

(b) Partial Restriction on Data Export

Under this category, copies of the data can leave the country. However, an updated version of the data must always reside within the host country. This variant is favoured by Indonesia and Malaysia.

(c) Consent-based Data Export

This variant of Data Localization mandates the foreign entities to seek consent from their local consumers before approving a cross-border data transfer. Brazil and Switzerland follow a variant of this category.

(d) Standards-based Data Export

Arguably, this is the least severe form of Data Localization. Under this category, foreign entities are allowed to transfer the "permitted" data across borders freely, provided that the standards prescribed by the host country for security and privacy are fully met.

Ever since "Edward Snowden, a former contractor with CIA, leaked to the media details of extensive Internet and phone surveillance by the American intelligence agency, in 2013,"[23] several governments have responded by way of stricter regulation of cross-border data flows. Most of the new laws are aimed at preventing sensitive data from leaving the domestic territory, with a view to protect themselves from foreign interception and surveillance.

Governments have viewed, for various reasons known to them, Data Localization as a means to ensure safety and protection of data in the event of a global geopolitical crisis. For instance, a significant amount of cross-border data flows move through undersea cables. This is viewed as a concern simply because the location of almost every undersea cable in the world is publicly known. While there is a broad acknowledgment that likelihood of such a situation arising is low, this concern has persisted. Moreover, Data Localization is also seen as a step to avoid the vulnerabilities arising from the fiber optic cable network.[24]

Governments have many reasons to believe that access to data is more secured when the data resides locally. This is important to note, especially in the context of civil and criminal investigations. Recently, in many countries issues such as unaccounted money, corruption, and high-profile banking scams have led to several governments exercising greater scrutiny, particularly of financial transactions by their citizens.

Yet another reason to justify Data Localization and unfettered access to data relates to effective law enforcement. By placing data servers inside a country's territory, it is apparently deemed to be easier for law enforcement agencies to gather evidence. Should the information be within a nation-state's jurisdiction, there is easier and faster access, in contrast to the tedious processes that law enforcement agencies might have to go through if and when the data is stored somewhere else in the world. Moreover, locating the servers domestically also means that the probability of foreign bodies declining access to the host country's own data, is considerably lower.

Not surprisingly, the number of Data Localization measures are on the rise globally. A telling statistic in this regard comes from the European Centre for International Political Economy (ECIPE). The think tank has indicated that in the decade ended in 2016, the number of significant Data Localization measures in the world's large economies increased almost threefold, from 31 to 84.[25]

On the other hand, relevant literature is full of studies that argue against Data Localization. For example, Bauer, Ferracane, Lee and Van der Marel study (2016) in the context of European

Union points out that "data regulations in general and Data Localization requirements, in particular, can take different forms according to stated objectives and this can affect downstream industries in many different economically distorting ways".[26]

Data Localization, it is argued, has the potential to threaten progress by hindering the development of goods and services and slowing economic growth. Increase in Data Localization laws, it is insisted, are likely to affect different sectors of the economy. According to a study conducted by a British think tank Chatham House and the Centre for International Governance Innovation (CIGI),[27] in the context of a group of emerging economies and the European Union, Data Localization laws have had serious adverse effects on the economies concerned. They have increased prices and lowered productivity, resulting in some loss of real GDP, annually. The Study has also quantified total factor productivity (TFP) costs that some countries have incurred due to the data flow regulations.

Champions of the free cross-border data flows claim that there is a lack of evidence that Data Localization requirements serve to improve data protection. On the contrary, they point out that there is a case to support the argument that localization can disrupt operations and will have a negative impact on any economy where it is implemented, as follows:

i.) In many cases, it is not possible to process all data locally and maintain the same quality of service as could otherwise be achieved (for example, round-the-clock, follow-the-sun customer service). This is essential for companies that operate globally and follow different markets.

ii.) Data Localization restrictions risk significantly impairing innovation by raising costs to potentially prohibitive levels for small and medium enterprises.[28]

iii.) Data Localization restrictions can seriously undermine a country's ability to leverage emerging revolutionary technologies that rely significantly on global networks, like Cloud computing and AI.

Thus, the adverse impact of Data Localization when forced on the industry, would include: economic cost of localization, organizational complexities that stem from it, impact on efficiency, reduction of the global footprint, reduced access to financial services and the challenges to global technology strategies.

The report presented by Chatham House and CIGI argues, and rightly so, that increasing organizational complexities does fundamentally alter business economics at a country or enterprise level. At the same time, there is no denying that the complexity of dealing with data location regulations would add another challenge for managers to overcome".[29]

Data Localization, it is argued, would cause a major rearrangement of data servers around the world. Whole countries and regions could potentially be out of the running to host data servers for global players. This could mean a severe blow to emerging market economies who have a very large and growing IT and service sectors.

In the ultimate analysis, it is up to the individual nation-state to take the call, not only in the interest of their economy and society, but also to keep themselves at the forefront of technological innovations which are changing the world for all times to come.

III. International Cooperation on Data Sharing

The global data governance ecosystem is evolving at a fast pace. As communities, societies and citizens increasingly migrate to digital and online platforms to carry out their personal and professional activities, it is clear that data is going to be at the center fold of all human activities in the times to come.

One area where greater emphasis needs to be placed is forming international alliances towards cooperation in data sharing. It may be recalled that international trade originated from universally recognized principles across-borders which were subsequently given the force of law by way of UN charters and international treaties.

Governments in Asia are working to develop and implement data privacy frameworks that can adequately protect the data of their citizens, while also allowing it to flow across-borders in ways that support trade and innovation. These frameworks encourage convergence across the region, which enables data to flow while maintaining a similar level of protection.

Across Asia, data privacy discussions are gaining traction, at both domestic as well as bilateral and multilateral level. This primarily includes ASEAN and APEC countries. The advantage of driving a privacy regime at a global level in Asia will help member countries' approach to data protection, reduce barriers to investments in the short and medium run along with a more specific compliance environment for businesses that wish to operate in that country.[30]

Asia is transitioning to be the growth engine of the world in the 21st century. Leveraging existing international mechanisms, such as the APEC's Cross-Border Privacy Rules, standard contractual clauses recognized by the European Commission, and binding corporate rules (recognized under the EU's GDPR) must be the way forward rather than creating national versions of these same mechanisms.

What is imperatively needed is developing a multilateral data-sharing framework that is based on the principles of adequacy and reciprocity.

Globally, countries are moving towards regional and multilateral data sharing and privacy frameworks that support cross-border data flow between those jurisdictions that deploy a high-standard of privacy regime in their national territories.

The OECD encourages countries to refrain from restricting data flows where there are sufficient safeguards and if the other country also observes similar guidelines. Similarly, the APEC provides that governments should ensure that there are no unreasonable impediments to cross-border data transfers while protecting the privacy and security of personal information. The ASEAN Framework objective is to "strengthen the protection of personal data in ASEAN and to facilitate cooperation among the Participants, intending to contributing to the promotion and growth of regional and global trade and the flow of information"[31] though it does not explicitly address restrictions on data flow.

However, the GDPR and to a lesser extent Convention 108 are more restrictive in the transfer of personal data of EU citizens to third countries or international organizations, unless they meet a set of conditions, ensure that the level of protection of individuals provided by the GDPR is not undermined.[32]

The conditions which are legally enforceable include the designation by the European Commission that a country meets an "adequate" level of personal data protection, or where standard contractual clauses or Binding Corporate Rules (BCRs) exist. It should be recognized, however, that the EU approach also prohibits restrictions on data flows between Member States/Parties.[33]

The primary issue in formulating a multilateral privacy framework for data sharing at the regional, continental or global level remains the same, i.e. alignment of a nation's right to territorial sovereignty with global cooperation.

A multilateral privacy framework assumes great significance on account of the following key attributes of Cloud data storage:

i.) Multiplicity - whereas physical information generally exists in one place at one time, data can be stored in multiple places at one time. Consequently, accessing the contents of a single, apparently unitary account may require retrieval of data stored on the territory of multiple sovereigns.

ii.) Attribution and Traceability – It's not mandatory or necessary to have any linkages or correspondence between the physical location of the data sets and the person to whom the data belongs.

iii.) Automation – since Cloud data is partitioned, stored, and moved automatically by algorithms, in several situations, it may be difficult for one to confirm with certainty the physical location of the data.

iv.) Anonymity – the identity, nationality, and residency of a data principle may not be known to the data processor or the data fiduciary.

Data in today's time and age is a commodity and consequently needs to be discussed not only at a bilateral level but at a multilateral level because the flow of data is seamless and without borders. Given the nature of technology and resultant blurring of territorial boundaries in case of data traffic and storage, it is essential for law enforcement agencies to have means and mechanisms for accessing the data and information beyond their territorial jurisdiction.

The current generation of Mutual Legal Assistance Treaties (MLATs), which were designed to facilitate the transfer of tangible evidence across borders, are functionally obsolete. The need of the hour is to have multilateral arrangements, specifically for data sharing, in light of the nature and attributes of digital data mentioned above and privacy and security issues. This will meet the local demands for law enforcement yet at the same time prevent forced localization measures.

Any multilateral data sharing framework must be based upon the following fundamental principles:

i.) Certainty and conclusiveness in the process for requesting and executing the data access and sharing requests.

ii.) Countries should cooperate on international security, counter-terrorism and countering money laundering, and this should be universally accepted and adopted grounds for data access.

iii.) Surveillance, other than preventive surveillance should be prohibited.

iv.) Reasons of refusal to grant access may be specified as a "negative list".

v.) Manner and procedure for functional implementation may be specified.

Several organizations and entities have done some seminal work in this field and have come up with interesting findings and propositions. Some such noteworthy projects are as follows:

i.) Findings of the Group of Governmental Experts on Developments in the Field of Information and Telecommunications in the Context of International Security.[34]

ii.) The MAPPING (Managing Alternatives for Privacy, Property and Internet Governance) Project.[35]

iii.) The Draft Legal Instrument on Government-Led Surveillance (LI).[36] The LI suggests the creation of an International Data Access Warrant (IDAW) that governments can use to obtain private data for investigative purposes. The current LI further suggests the creation of an International Data Access Authority composed of retired judges from the contracting states who would evaluate IDAW applications against international human rights norms before authorizing them.

IV. International Cooperation Issues on Privacy

Random data sets do not make sense to humans, however, feeding such data sets to AI algorithms can help in identifying users who generate this data without compelling them to provide their names.

Since AI can compare two or more databases, it may not need the name and/or social security numbers to identify individuals.

Data sets that contain "location stamps" and time stamps can help track the trend of a person's mobility.

There are two issues here. First, how can AI improve privacy mechanisms for companies handling large amount of data sets. And second, how can privacy be incorporated into the AI algorithms.

AI can support companies to identify and respond in real time to individuals to prevent inappropriate usage of personal information or theft of data. Moreover, AI can also support the development of privacy tools. One example is 'Polisis' that stands for 'Privacy Policy Analysis'. This application reads privacy policies and develops insights, for example, an executive summary or a flow chart of the kind of data collected and its recipient. Additionally, it also informs if the consumer can opt out of such data collection or sharing.

Data is driving the economies of the present, and will do so in the future as well, and in such an ecosystem, the idea of Privacy by Design becomes all important to protect the interests and rights of citizens.

Privacy by Design techniques should be embedded into AI development. With the rise of data collection and storage, notions of consent and privacy notices would have to be reconsidered.

"The model of 'clickwrap' contracts, which allows the user to click on 'I accept' button without reading a long, verbose and unintelligible privacy terms and conditions needs to be also revisited. Privacy by Design techniques can be incorporated not only at the level of privacy notices but also at each level of information flow till its storage and processing stage."[37]

Further notions of transparency, accountability and fairness must be incorporated. While there can be no strict set of rules or policy guidelines, which can bind an algorithm designer, but, best practices following constitutional standards jurisdiction-wise can be developed as a benchmark.

Some of the techniques to enhance privacy of existing AI tools comprise "differential privacy", "homomorphic encryption" and "generative adversarial networks". In addition to this, certifications of schemes and privacy seals can also be considered, which help demonstrate the compliance level deployed by organizations.

Privacy and AI not only can complement each other but also enhance the overall output while at the same time provide consumers the best user experience and simultaneously protect their data.

Privacy by Design works when a risk is identified, looked upon as an opportunity towards finding creative solutions to secure data. This is a requirement nowadays as AI is proliferating fast. The opportunity therefore is to not just solve problems but to do it in a sustainable manner for protecting digital rights of the people.

V. Banking-Finance 4.0: Regulatory Dimensions

With emerging technologies, evidently, there is a need for a new approach by central banks across the world to regulate the so-called Banking-Finance 4.0. Advanced economies, as well as several emerging market economies have elaborate and rigid regulatory frameworks existing from the time when everything was paper-based, data was insufficient and computing was expensive. But now, both data and computing are ubiquitous and cheap. Brett King, therefore, suggests that we need to create a whole new model for this digital age, not only to regulate the digital markets but also to deploy new technologies in the regulatory process.[38]

Change will not be easy for central banks. As is well known, institutions often get entangled in their legacy products and practices that have worked for them for long. Therefore, they become "vulnerable to displacement by superior disruptive technologies".[39] As regulators, more often than not, Central banks are bound by their well-founded and constrained set of systems in which they operate. They are simply not cut out for these metamorphic changes. As Sheng (2018) puts it, "if the banks are behind the curve in terms of technology, the regulators are even further behind, since they learn mostly from those whom they regulate".[40]

Brett King argues that changes in regulation require both an identification of a system change and changes in law or operational structure to implement.[41] Moreover, since the regulators are not generally accused of being innovators, they are likely to respond to innovations as a systemic risk.

A PricewaterhouseCoopers (PwC) Report (2016) observes:

For central banks there is the challenge of overseeing ever extending industry boundaries in areas such as marketplace lending, new digital currencies, shadow banking and business models using block-chain technology. These developments are creating real-time, decentralized networks and potentially redefining payments transactions and trading. Central banks need new skills and systems to keep pace with the complexity of evolving markets. As supervision evolves from historical to predictive tools, large investments will be needed to manage "big data". They will need data analytics tools to identify where risks are building up in the financial system, such as in the pricing of risk, cross-border risks, data quality and cybercrime.[42]

According to Brett King, "another key challenge is that regulators must join the global technology shift to cloud computing. Central bankers are generally reluctant to adoption of cloud-based systems due to security concerns. However, cloud systems services are actually far more secure, not less. In cloud services the data is all online, which means it is protected from physical disasters like fire, and it can be secured efficiently".[43]

Dong He (2018) rightly points out that "there are both challenges and opportunities for central banks in the digital age. Central banks must maintain the public's trust in fiat currencies and participate in a digital, sharing, and decentralized service economy. According to him, Central banks can remain relevant by providing more stable units of account than crypto assets and by making central bank money attractive as a medium of exchange in the digital economy".[44]

The PwC Report cited earlier emphasizes that Central banks are going to play an important role to facilitate the public debate over what society wants from the banking system under the emerging scenario on Banking-Finance 4.0. The debate will also determine the Central bank's role in the regulation of the future financial system and wider economy. Questions include the extent to which the Central banks are responsible for ensuring that credit goes to the productive sectors of the economy and what actions it can take if this comes up against public or political opposition. "There would also need to be clarity over defined policy areas such as inflation targets or measures of stability, against which progress and success could be judged".[45]

The problem really lies with the approach of Central banks towards the emerging transformative technologies. If Central banks remain committed to their traditional way of functioning and do not want any hindrance whatsoever in their regulation, they are likely to face obsolescence in none too distant future.

Christine Lagarde, IMF Managing Director, in one of her speeches (September 2017), titled "Central Banking and FinTech – A Brave New World?" had discussed how FinTechs will change the Central banking over the next generations.

While referring to formulation of monetary policy in the context of the new models of financial intermediation engendered by the emerging New-Age technology, she said:

"Today's central banks typically affect asset prices (i.e. interest rates) through primary dealers, or big banks, to whom they provide liquidity at fixed prices – the so called open market operations. But if these banks were to become less relevant in the new financial world, and demand for Central bank balances were to diminish, could monetary policy transmission remain as effective?"

"If anything, Central banks may well have to increase the number of counterparties to their operations.... More counterparties imply more firms falling under the 'Central banks' regulatory umbrella – which is the price to pay for liquidity on a rainy day. Whether the future holds more or less rain is an open question. Still, better regulating shadow banks looks all the more pressing."

Lagarde rightly insisted that Central banks would have to "expand their focus, from financial entities to financial activities". While referring to virtual currencies, she argued that "it may not be wise to dismiss virtual currencies" because many of the ongoing technology challenges could be addressed over time. As to the question, why might citizens hold virtual currencies rather than physical dollars, euros, or sterling, she argued that it "may one day be easier and safer than obtaining paper bills, especially in the remote regions. And because virtual currencies could actually become more stable", she also observed that "if privately issued virtual currencies remain risky and unstable, citizens may even call on Central banks to provide digital forms of legal tender".

In essence, her argument is that: "Virtual currencies might just give existing currencies and monetary policy a run for their money. The best response by Central bankers is to continue running effective monetary policy, while being open to fresh ideas and new demands, as economies evolve".[46]

With the emergence of cryptocurrencies Central bankers are looking for ways to ensure the integrity of markets and payment systems, to protect consumers and investors, and to safeguard overall financial stability.[47] An important challenge is to combat illicit usage of funds. At the same time, Central banks would want to preserve long-run incentives for innovation. A related question is whether Central banks should issue their own bank digital currency.

Claire Wilson argues that "the current situation in Venezuela and the United Arab Emirates show that state-issued cryptocurrencies and private cryptocurrencies can operate side by side". However, it may be noted that State-backed cryptocurrencies have only started to emerge this year. It is perhaps too early to be able to ascertain whether they do pose a serious threat to the future of privately issued cryptocurrencies".

There are key regulatory challenges for the government and the regulatory authorities. First is that of anti-money laundering measures and secondly, combating the financing of terrorism. Due to anonymity, it is difficult to quantify the extent to which cryptocurrencies are being used. It is hard to track the receiver and sender of the currency. This could be the hub of illicit activities.[48]

Yet another challenge is securing rules and regulations and ensuring consumer protection. One common issue with digital currency is digital theft. Due to the emergence of Bitcoin, ransomware started growing globally, when the only way to release files held ransom was to send Bitcoins to the hackers.

A long-term challenge for consideration is that of stability of financial systems. The large scale and widespread use of it is yet to be seen. The use of cryptocurrencies will definitely give rise to financial vulnerabilities and systematic risks.[49]

Agustin Carstens, General Manager of the Bank for International Settlement (BIS) suggested recently (March 2019) the following course of action: In order to move forward, Central banks need to consider at least four important aspects:

i.) First, they must develop a fuller understanding of the FinTech business in their jurisdiction.

ii.) Second, careful thought should be given to adjusting the regulatory perimeter so as to accommodate new entities or activities and ensure adequate control of risks and a level playing field. Activity-based regulation could be more a complement than a substitute for the traditional entity-based regulation.

iii.) Third, regulatory actions should be coordinated at the global level to the extent possible. Financial innovations are a global phenomenon given that providers of these new services and products often act on a cross-border basis, taking advantage of the new virtual distribution channels.

iv.) Fourth, regulators should also embrace technology, such as AI and its practical application in machine learning.[50]

In her most recent address (June 8, 2019), Christine Lagarde lauded the impact of Banking-Finance 4.0 on financial inclusion. She pointed out that "the rapid growth of FinTech has promoted access to cheap payment and settlement systems for low-income groups in developing countries, where traditional banking networks are scarce".[51]

In the same address, she observed that "global policy makers are wary of FinTechs' increasing dominance in mobile payments, as they rethink ways to regulate the banking system. They fear that the stability of the whole economic system might be at risk, as FinTechs threaten to disrupt the centuries-old financial industry".[52]

On behalf of the International Monetary Fund (IMF) which she heads, she urged the "world leaders to take more action to address the increased use of FinTech, which, by controlling payments and settlement systems, could make the international financial networks vulnerable".

According to her, "This presents a unique systematic challenge to financial stability and efficiency, and one I hope we can touch on during the G-20, and address in a cooperative and consistent fashion".[53]

A Summing Up

The preceding discussion on regulation and strategic development of the New-Age Technology underlying the Industrial Revolution 4.0 can best be summarized, as formulated in the Group of Twenty (G-20) Osaka Leaders' Declaration (Para 12) (June 29, 2019):[54]

> To further promote innovation in the digital economy, we support the sharing of good practices on effective policy and regulatory approaches and frameworks that are innovative as well as agile, flexible, and adapted to the digital era, including through the use of regulatory sandboxes. The responsible development and use of Artificial Intelligence (AI) can be a driving force to help advance the SDGs and to realize a sustainable and inclusive society. To foster public trust and confidence in AI technologies and fully realize their potential, we commit to a human-centred approach to AI, and welcome the non-binding G20 AI Principles, drawn from the Organization for Economic Cooperation and Development (OECD) recommendation on AI. Further, we recognize the growing importance of promoting security in the digital economy and of addressing security gaps and vulnerabilities. We affirm the importance of protection of intellectual property. Along with the rapid expansion of emerging technologies including the Internet of Things (IoT), the value of an ongoing discussion on security in the digital economy is growing. We, as G20 members, affirm the need to further work on these urgent challenges. We reaffirm the importance of bridging the digital divide and fostering the adoption of digitalization among micro, small and medium enterprises (MSMEs) and all individuals, particularly vulnerable groups and also encourage networking and experience-sharing among cities for the development of smart cities.

PART • FOUR

A CASE-STUDY

Chapter 14

New-Age Technology:
An Overview of Indian Experience

14

New-Age Technology: An Overview of Indian Experience

"We will… lay focus on new-age skills like Artificial Intelligence (AI), Internet of Things, Big Data, 3D Printing, Virtual Reality and Robotics, which are valued highly both within and outside the country,…"

Nirmala Sitharaman[1]
Minister of Finance, India

Introduction

Recent technological advances are already having a profound effect on everything in India. Our life, our work, even our value systems are being influenced; our economy and society are beginning to transform. And this is happening in almost every sector – education, health, finance, and governance. Even the attitudes, behavior and culture are being shaped in India by the New-Age Technology.[2]

This final chapter has been organized as follows:

Section I presents a snapshot of the prevailing development and application of the New-Age Technology in India. Against that background, Section II offers a perspective on India's policy in respect of AI, Big Data, Privacy, Cybersecurity and Surveillance.

I. Emergence of New-Age Technology in India: A Snapshot

A. Technology and Education

Recent technological advances are already being a game changer in all facets of education in India. With increasingly enhancing communication, both the speed of generating research-based knowledge as well as the pace of its dissemination far and wide have been rising sharply. Consequently, issues relating to the delivery of high-quality education—when, where, how, in what form and by whom—are being resolved at a fraction of costs associated with the traditional 'brick and mortar' approach.

Illustratively, the Massive Online Open Courses (MOOCs) today signify high-quality courses, created by top experts in respective diverse fields, made available virtually free of cost, anywhere in the country to any person who has Internet connectivity.

B. Technology and Health

In developing countries like India which more often than not, suffer from acute shortage of not only specialists but also of qualified paramedics, especially in the rural areas, telemedicine has been offering an effective intermediate solution.

Reportedly, the Indian Space Research Organization (ISRO) has recently developed and deployed a telemedicine facility in Siachen Glacier (in the Eastern Himalayan region of Jammu and Kashmir where the Line of Control between India and Pakistan ends). This communication through satellite-enabled telemedicine nodes is meant for improving emergency medical support to soldiers posted in high altitude areas and is expected to be a paradigm shift in the delivery of life-saving health care.[3]

C. Technology, Money and Finance

In India, digital money was already making its presence felt when digital payments received a big boost in November 2016. In a surprise move to contain the black money (i.e. unaccounted money believed to be primarily held in currency hoards), high-denomination currency notes (Rs 500 and Rs 1000) were demonetized. This shock accelerated India's progression towards less-cash economy.

Today, e-commerce is booming in India. Any individual can surf the relevant website to order merchandise from elsewhere, within or outside India.

As Karnik (2018) puts it, for this purpose there is a "broader trend: to shift all complexity to the back-end and make the user's experience as easy and comfortable as possible... there are mobile phone apps used for complex tasks, which can be used even by a semi-literate, "non-tech" person. One example is payment apps..."[4]

In India, point of sale (PoS) machines for using credit or debit cards are available only at big establishments. Under the circumstances, mobile phones payment apps have emerged as a very useful option. Preloaded money in one phone's app could be readily transferred to another, through just a few easy steps. The ease and user-friendliness of the payment process has made it very popular. The Government of India has facilitated these transactions through mechanism such as Bharat Interface for Money (BHIM) app.

D. Technology and Digital Payments in India

Financial Technology is rapidly evolving and spawning new businesses and new sectors as a whole. In India, this is happening primarily due to increased mobile penetration and increased Internet speed. Faster digital transactions coupled with incentives like cashbacks and deep discounts are creating a sense of reliability and security amongst customers.

The value of the BHIM Unified Payments Interface (UPI)—one of the Digital Payments mechanisms—transactions skyrocketed to Rs 1,335 billion by March, 2019 while the volume of transactions reached Rs 799 million, up from a meagre Rs 178 million in March 2018.[5] The digital payment industry is gaining momentum and is projected to grow at an exponential rate of 81 per cent since the existing digital payment users prefer the medium over other non-cash payment methods like cheques or demand drafts.[6]

According to a 2018 report by investment banking firm, Credit Suisse, the Indian digital payments industry which is currently worth around US$ 200 billion, is expected to grow five-fold to reach US$ 1 trillion by 2023.[7] Apart from UPIs, the digital payment industry growth will also be led by digital/mobile wallets.

E. Technology and Agriculture

In the agriculture sector, India has utilized the AI capabilities in addressing several problems. For example, India has created a database of all plant diseases. This allows farmers to protect their crops using image recognition and AI. Microsoft Cortana along with the International Crops Research Institute for the Semi-Arid Tropics (ICRISAT) has also developed an app which advises farmers regarding the optimal date for sowing seeds.[8]

II. New-Age Technology and Policy Issues in India

A. Policy on Artificial Intelligence (AI)

The National Institution for Transforming India (NITI Aayog), the think tank that replaced the erstwhile Planning Commission, came up with India's AI Policy through a Working Paper titled, "National Strategy for Artificial Intelligence: #AIforAll".

The NITI Aayog Working Paper identifies five focus areas where AI development could enable both growth and greater inclusion: education, healthcare, agriculture, urban-/smart-city infrastructure, and transportation and mobility. It also discusses five barriers that need to be overcome: limited research expertise, absence of enabling data ecosystems, high resource cost and lack of awareness for adoption, paucity of regulations around privacy and security, and non-existence of a collaborative approach to adoption and applications.

The Working Paper proposes a two-pronged approach for promoting AI research: First, creation of Centres of Research Excellence in AI (COREs), which will be academic research hubs, and secondly, International Centres for Transformational Artificial Intelligence, which will be led by the industry.

The NITI Aayog Working Paper includes several recommendations, but has been deliberately couched in terms of an 'essential pre-read' intended merely to begin the discussion for what will be an evolving National Strategy for Artificial Intelligence.

B. Technology, Big Data and Privacy Policy in India

In India, the Department of Science and Technology (DST), under the Ministry of Science and Technology and Earth Sciences has been entrusted with the task of developing Big Data Analytics (BDA) ecosystem. Big Data Analytics infrastructure development in India is being steered by the C-DAC (Centre for Development of Advanced Computing), Ministry of Electronics and Information Technology. C-DAC has already created a state-of-the-art hardware system and networking environment at its various facilities. As more and more data are generated with phones, laptops and other Internet of Things, it is imperative that the Big Data is stored somewhere securedly and analyzed independently.

The Unique Identification Number (UIN) or "Aadhaar" in India allows the Government to verify the identity of Indian citizens. This can be very useful for planning social programs.

The Aadhaar and Other Laws (Amendments) Bill, 2019 was passed by the Indian Parliament on 8 July 2019. This Act replaces an Ordinance promulgated earlier on 2 March 2019.

Under the Aadhaar Act, an individual's identity may be verified by Aadhaar "authentication". Aadhaar Act allows the verification of an individual's identity, as defined under the Act, which involves submitting the Aadhaar number, and their biometric or demographic information to the Central Identities Data Repository for verification.

The amended Act additionally allows "offline verification" of an individual's identity, without authentication, through modes specified by the Unique Identification Authority of India (UIDAI) by regulations.

During offline verification, the agency must (i) obtain the consent of the individual, (ii) inform them of alternatives to sharing information, and (iii) not collect, use or store Aadhaar number or biometric information.

The amended Act states that an individual may voluntarily use his or her Aadhaar number to establish his or her identity, by authentication or offline verification. The revised Act also states that the authentication of an individual's identity *via* Aadhaar, for the provision of any service, may be made mandatory only by a law of Parliament.

In India, a sound strategy so as to protect such "Big Data" is still evolving. The Government is planning to set up a Data Protection Authority to probe any "data security breach".

In India, most of the contextualization around privacy has originated from English common law. However, the notions of privacy are pre-dated to Hindu and Islamic legal systems that have their own rich jurisprudence. In the modern era, India not only has declared explicitly that Right to Privacy is a fundamental right, but the country is also committed to a global human rights regime.[9]

In the wake of the recognition of the Right to Privacy as a fundamental right, the Ministry of Electronics and Information Technology came up with a draft *Personal Data Protection Bill, 2018.* The main purpose of this draft bill is to lay down policies and procedures seeking to minimize the intrusion into the privacy of individuals caused by the collection and usage of their personal data. The Bill is yet to be passed by the Indian Parliament.

C. Technology and Cybersecurity Policy in India

With the advent of the New-Age Technology, there could possibly be instances where these emerging technologies fall into unwanted hands.

In 2008, the Indian Information Technology Act was reworked and among several amendments, Section 70 was introduced which defined Critical Information Infrastructure (CII) as:

> *"those facilities, systems or functions whose incapacity or destruction would cause a debilitating impact on national security, governance, economy and social well-being of a nation."*

The emergence of new non-state actors post-Cold War era in the form of terrorist organizations has dramatically changed the national security environment for India.

Cyberattacks done on the CII can cause huge financial losses, environmental damage and pose a threat to defense, security and integrity of India.[10]

The Government of India has designated the National Critical Information Infrastructure Protection Centre (NCIIPC) of National Technical Research Organization (NTRO) as the nodal agency under Section 70A (1) of the Information Technology (Amendment) Act 2008 for taking all measures, including associated research and development for the protection of CIIs in India.

India has introduced its first National Cybersecurity Policy in 2013, which is currently under revision.

D. Technology and Surveillance Policy in India

The current surveillance laws in India are mostly remnants from India's pre-technological colonial past. In the age of rapid technological growth, the presence of mobile apps and new communication devices, digital surveillance has extended to the interception of data, along with the interception of calls. New surveillance reform calls for an extensive focus on data privacy. While the Indian Telegraph Act, 1885 deals with the interception of calls, the Information Technology Act, 2000 deals with data.

In sum, India is ushering into an era of high productivity where technology will play a key role in facilitating and sustaining economic growth and development. Larger public investment in AI research as well as in skill development commensurate with New-Age Technology, and a progressive privacy law based on the principles enshrined in the Supreme Court judgment of "Right to Privacy", are the need of the hour. For India, it is crucial to balance the need to deploy technology for development, while at the same time protecting the privacy and security of millions of citizens. Moreover, law enforcement requirements of the Government must not, and cannot be compromised. At the same time, due process of law for data access in India needs to be formulated, which is grounded in judicial safeguards and Parliamentary oversight. Privacy is not an impediment to growth, but an enabler in a digital economy, where technology will become even more pervasive in the future.

NOTES

Chapter 1

1. Wilson, E.O. (2012). *The Social Conquest of Earth,* New York, London: Liveright, W.W. Norton & Company.
2. Gibbs, S. (July 27, 2015). "Musk, Wozniak and Hawking urge ban on warfare AI and Autonomous Weapons*", The Guardian*, retrieved from: https://www.theguardian.com/technology/2015/jul/27/musk-wozniak-hawking-ban-ai-autonomous-weapons (accessed February 03, 2019).
3. Gibbs, S. (October 27, 2014). "Elon Musk: Artificial Intelligence is our biggest existential threat"*, The Guardian,* retrieved from: https://www.theguardian.com/technology/2014/oct/27/elon-musk-artificial-intelligence-ai-biggest-existential-threat (accessed February 03, 2019).
4. Morgan, L.H. (1877). *Ancient Society*, Chicago: Charles H. Kerr Publishing Company.
5. White, L.A. (1949). *The Science of Culture: A Study of Man and Civilization*, Percheron Press/Eliot Werner Publications.
6. Lenski, G. (2005). *Ecological-Evolutionary Theory: Principle and Applications*, Paradigm Publishers -2005/Routledge – 2016.
7. In the late 1970s, Alvin Toffler (author of *Future Shock*), among others, argued that the era of industrial society is coming to an end.
8. Schwab, K. (2017). "The Fourth Industrial Revolution: what it means, how to respond", *World Economic Forum*, retrieved From: https://www.weforum.org/agenda/2016/01/the-fourth-industrial-revolution-what-it-means-and-how-to-respond/
9. Mashelkar, Dr R. (2017), "Exponential Technology, Industry 4.0 and Future of Jobs", Paper presented at Pune International Centre (PIC) Conference on the theme "Accelerating Growth through Employment", Pune, December 8-9, 2017.
10. Pramanik, B. (2018), "From Asimov to Harari – The Ethics of Analytics", Keynote Address delivered at the Goa Institute of Management, Bengaluru, September 22, 2018.
11. "What is the Fourth Industrial Revolution?", *World Economic Forum (WEF).*
12. See Pramanik, B. (2018), *op. cit.*, as well as Karnik, K., (2018), *Evolution Decoding India's Disruptive Tech story*, Rupa Publications India.

Chapter 2

1. Both quotes are from www.psychologytoday.com (accessed May 1, 2019).
2. Howes, Anton. (October 22, 2016). "Is Innovation in Human Nature?" *Medium*, retrieved from: https://medium.com/@antonhowes/is-innovation-in-human-nature-48c2578e27ba (accessed February 05, 2019).
3. As Stephen Hawking said: "The development of full artificial intelligence could spell the end of the human race…..It would take off on its own, and re-design itself at an ever increasing rate. Humans who are limited by slow biological evolution, couldn't complete, and would be superseded".
4. Wilson, H. J. & Daugherty, P. R. (April 04, 2019). "How Humans and AI Are Working Together in 1,500 Companies", *Harvard Business Review*, retrieved from https://hbr.org/2018/07/collaborative-intelligence-humans-and-ai-are-joining-forces.
5. Vincent, J. (17 Jul. 2017). "Elon Musk says we need to regulate AI before it becomes a danger to humanity", *The Verge*, Retrieved from: https://www.theverge.com/2017/7/17/15980954/elon-musk-ai-regulation-existential-threat (accessed May 2, 2019).
6. Buchanan, B.G. (December 05, 2005). "A (Very) Brief History of Artificial Intelligence", *AI Magazine,* 26(4).
7. Ray, S. (August 11, 2018). "History of AI", *Medium*, retrieved from https://towardsdatascience.com/history-of-ai-484a86fc16ef (accessed May 2, 2019).
8. Stanford University's Study Panel's Report on "100 Years of AI", (2016), p. 12.
9. Yudkowsky, E. (2008). "Artificial Intelligence as a Positive and Negative Factor in Global Risk." in *Global Catastrophic Risks*, eds. Nick Bostrom and Milan M. Ćirković, pp. 305-45, New York: Oxford University Press.
10. Nilsson, N.J. (2010). *The Quest for Artificial Intelligence: A History of Ideas and Achievements*, Cambridge, UK: Cambridge University Press.
11. McCorduck, P. *Machines Who Think: A Personal Inquiry into the History and Prospects of Artificial Intelligence,* 2nd ed., Natick, Massachusetts: A. K. Peters, Ltd., 2004; San Francisco: W. H. Freeman, 1979. Citations are to the Peters edition.
12. Woodford, C., (March 30, 2018). "The odd paradox in defining artificial intelligence", *Medium*, Retrieved From: https://medium.com/machines-learning/the-odd-paradox-in-defining-artificial-intelligence-fc62b18dbdd0 (accessed May 2, 2019).
13. Cohen, M.R., & Nagel, E. (2002). *An Introduction to Logic and Scientific Method.* Safety Harbor, Florida: Simon Publications.
14. Britannica, T.E. (December 23, 2013). "Syllogism", *Encyclopaedia Britannica,* retrieved from: https://www.britannica.com/topic/syllogism (accessed May 2, 2019).

15. For example, see the proof of Pythagorean Theorem illustrated in Annexure I.
16. Hawthorne, J. (March 19, 2018). "Inductive Logic", *Stanford Encyclopedia of Philosophy,* retrieved from: https://plato.stanford.edu/entries/logic-inductive/ (accessed May 2, 2019).
17. Henderson, L. (March 21, 2018). "The Problem of Induction", *Stanford Encyclopedia of Philosophy,* retrieved from: https://plato.stanford.edu/entries/induction-problem/ (accessed May 2, 2019).
18. Hurwitz, J., & Kirsch, D. (2018). *Machine Learning for Dummies*, Hoboken, New Jersey: John Wiley & Sons.
19. McNulty, E. (January 08, 2015). "What's The Difference Between Supervised and Unsupervised Learning?" *Dataconomy*, retrieved from: https://dataconomy.com/2015/01/whats-the-difference-between-supervised-and-unsupervised-learning/ (accessed May 2, 2019).
20. Marr, B. (October 01, 2018). "What Is Deep Learning AI? A Simple Guide with 8 Practical Examples", *Forbes*, retrieved from: https://www.forbes.com/sites/bernardmarr/2018/10/01/what-is-deep-learning-ai-a-simple-guide-with-8-practical-examples/#6f23af378d4b (accessed May 2, 2019).
21. Strogatz, S. (December 26, 2018). "One Giant Step for a Chess-Playing Machine", *The New York Times*, retrieved from: https://www.nytimes.com/2018/12/26/science/chess-artificial-intelligence.html (accessed May 2, 2019).
22. Coppey, L. (March 15, 2018). "What does AlphaGo vs Lee Sedol tell us about the interaction between humans and intelligent systems?" *Medium*, retrieved from: https://medium.com/point-nine-news/what-does-alphago-vs-8dadec65aaf (accessed May 2, 2019).
23. "Future of Artificial Intelligence Economic Growth" (n.d.), *Accenture*, retrieved from: https://www.accenture.com/us-en/insight-artificial-intelligence-future-growth (accessed May 2, 2019).
24. "Why Is Artificial Intelligence The Need To The Hour In The Manufacturing Industry?" (March 22, 2018), *Infiniti Research*, retrieved from: https://www.infinitiresearch.com/thoughts/artificial-intelligence-manufacturing-industry.
25. From ca-biomed.org (accessed May 2, 2019).
26. Al-Lazikani, B. (November 12, 2013). "Artificial intelligence uses biggest disease database to fight cancer", *The Conversation*, retrieved from https://theconversation.com/artificial-intelligence-uses-biggest-disease-database-to-fight-cancer-20050 (accessed February 06, 2019).
27. Chatbots are basically computer programs that conduct conversations via auditory or textual methods.

28. "Chatbots, A Game Changer for Banking & Healthcare, Saving $8 billion Annually by 2022" (9th May, 2017), *Juniper Research*, retrieved from: https://www.juniperresearch.com/press/press-releases/chatbots-a-game-changer-for-banking-healthcare (accessed February 06, 2019).

29. Bouman, K. (April 28, 2017). "How to take a picture of a black hole", *TED*, retrieved from: https://www.youtube.com/watch?v=BIvezCVcsYs (accessed February 06, 2019).

30. Alexa, (September 02, 2018). "Racial Bias and Gender Bias Examples in AI systems", *Medium*, retrieved from: https://medium.com/thoughts-and-reflections/racial-bias-and-gender-bias-examples-in-ai-systems-7211e4c166a1 (accessed February 06, 2019).

31. Dastin, J. (October 10, 2018). "Insight - Amazon scraps secret AI recruiting tool that showed bias against women", *Thomson Reuters*, retrieved from: https://in.reuters.com/article/amazon-com-jobs-automation/insight-amazon-scraps-secret-ai-recruiting-tool-that-showed-bias-against-women-idINKCN1MK0AH. (accessed February 06, 2019).

32. Rigby, M. J. (February 01, 2019). "Ethical Dimensions of Using Artificial Intelligence in Health Care", *AMA Journal of Ethics*, retrieved from: *https://journalofethics.ama-assn.org/article/ethical-dimensions-using-artificial-intelligence-health-care/2019-02.*

33. Basu, A. (October 23, 2018). "Discrimination in the Age of Artificial Intelligence", *Oxford Human Rights Hub,* retrieved from http://ohrh.law.ox.ac.uk/discrimination-in-the-age-of-artificial-intelligence.

34. "Ethics and autonomous weapon systems: An ethical basis for human control?" (April 3, 2018), *International Committee of the Red Cross*, retrieved from: https://www.icrc.org/en/document/ethics-and-autonomous-weapon-systems-ethical-basis-human-control (accessed February 06, 2019).

35. Downloaded from webhome.Auburn.edu (accessed May 26, 2019).

36. Paulas, R. (April 02, 2017). "How humans will lose control of artificial intelligence", *The Week*, retrieved from: https://theweek.com/articles/689359/how-humans-lose-control-artificial-intelligence (accessed February 06, 2019.

37. Good, I.J. (1965, January 01). "Speculations concerning the first ultra-intelligent machine", A*dvances in Computers,* Academic Press, retrieved from: https://philpapers.org/rec/GOOSCT (accessed February 06, 2019).

38. Creighton, J. (February 14, 2018). "The "father of artificial intelligence" says singularity is 30 years away", *Futurism*, retrieved from: https://futurism.com/father-artificial-intelligence-singularity-decades-away (accessed February 06, 2019)

39. Martin, S. (April 10, 2019). "Elon Musk WARNING: Artificial Intelligence could be an 'IMMORTAL DICTATOR"', *Express,* retrieved

from: https://www.express.co.uk/news/science/1112515/elon-musk-artificial-intelligence-ai-news-google-deepmind (accessed February 06, 2019)

40. "Proof Pythagorean theorem" (n.d.), *Pythagoras Theorem,* retrieved from http://pythagoras.nu/proof-pythagorean-theorem/

Chapter 3

1. This quote appears again and again from Sherlock Holmes, a fictional private detective created by Sir Arthur Conan Doyle. This and other quotes here have been taken from various collections readily available on the Internet
2. "Big Data Analytics: What it is and Why it Matters", *SAS,* retrieved from: https://www.sas.com/en_us/insights/analytics/big-data-analytics.html. Accessed 15 Nov. 2018 (accessed October 26, 2018)
3. Snjders, C., Matzat, U. and Reips, U. (2012), "Big Data: Big Gaps of Knowledge in the Field of Internet," *International Journal of Internet Science,* 7, pp 1-5.
4. Hilbert, M., Lopez, P. (April 01, 2011). "The World Technological Capacity to Store, Communicate, and Compute Information" *Science,* 6025(332).
5. Science Daily, May 22, 2013 (accessed October 26, 2018).
6. *Forbes*, May 21, 2018 (accessed October 26, 2018).
7. De Mairo, A., Greco M. and Grimal di, M. (2016), "A Formal Definition of Big Data Based on its Essential Features", *Library Review*, 65, 3, pp 122-135.
8. "Data, Data Everywhere" (February 25, 2010), *The Economist*, retrieved from: https://www.emc.com/collateral/analyst-reports/ar-the-economist-data-data-everywhere.pdf (accessed October 26, 2018.
9. "Big Data for Development: Opportunities and Challenges" (2012), *UN Global Pulse*, retrieved from: http://www.unglobalpulse.org/sites/default/files/BigDataforDevelopment-UNGlobalPulseJune2012.pdf.
10. "Big Data for Development: From Information to Knowledge Societies" (January 15, 2013), *Social Science Research Network (SSRN).*
11. "New Themes in Innovation Policy- New Requirements for Innovation Policy" (2013), *INNO-Grips Policy Brief No. 6*, retrieved from: http://innogrips.empirica.biz/fileadmin/innogrips/documents/01_policy%20 briefs/ig_policybrief_6_new-themes.pdf (accessed October 26, 2018).
12. *The Guardian*, July 3, 2018 (accessed November 1, 2018).
13. Cisco (2016), "Cisco Visual Networking Index: Complete Forecast Highlights", 2015-2020, as quoted in US Department of Commerce Report: *Measuring the Value of Cross-Border Data Flows*, September 2016.
14. Ibid, p. 32.

15. Lund, S. and Manyika, J. (April 20, 2017). "Defending Digital Globalization," *Foreign Affairs*, retrieved from: https://www.foreignaffairs.com/articles/world/2017-04-20/defending-digital-globalization (accessed November 1, 2018).
16. *World Bank Document: Open Knowledge Depository,* (accessed October 26, 2018).
17. As discussed in detail in US Department of Commerce Report (2017), cited earlier.
18. Lund and Manyika (2017), *op. cit.*
19. Ibid.
20. UNCTAD, 2015 estimate quoted in the US Department of commerce Report (2016).
21. Grimm. A. (May 5, 2016), "Trends in US Trade in ICT Services and in ICT-enabled Services", *Bureau of Economic Analysis*, retrieved from: https://apps.bea.gov/scb/pdf/2016/05%20May/0516_trends_%20in_us_trade_in_ict_serivces2.pdf
22. Nicholson. J, (2016), as quoted in the US Department of commerce Report, cited above.
23. US Department of Commerce Report, *op. cit.*
24. Lund and Manyika (2017), *op. cit.*
25. *Businesswire*, April 16, 2015 (accessed October 26, 2018).
26. Lund and Manyika (2017), *op. cit.*
27. Manyika, J., et al. (2011). "Big Data: The next frontier for innovation, competition, and productivity", *McKinsey Digital*, retrieved from: https://www.mckinsey.com/business-functions/digital-mckinsey/our-insights/big-data-the-next-frontier-for-innovation (accessed October 26, 2018).
28. G-20 Declaration (June 29, 2019).

Chapter 4

1. James Scott Brown – American author, formerly associated with the Institute of Critical Infrastructure Technology, a Washington DC based think tank.
2. MacDonald. K. (June 27, 2018). "Being Human: How Realistic Do We Want Robots to be?" *The Guardian*, retrieved from: https://www.theguardian.com/technology/2018/jun/27/being-human-realistic-robots-google-assistant-androids (accessed May 01, 2019).
3. Ibid.
4. Pinto, J. "Robotics Technology Trends: The Future of Robots by Jim Pinto", *Automation*, retrieved from: https://www.automation.com/library/articles-white-papers/articles-by-jim-pinto/robotics-technology-trends (accessed May 01, 2019).

5. Jim Pinto, Ibid.
6. Kaeser, J. (September 14, 2017). "Why Robots Won't Steal Human Jobs In Manufacturing", *Time*, retrieved from: http://time.com/4940374/joe-kaeser-siemens-robots-jobs/ (accessed May 01, 2019).
7. "NASA - People Are Robots, Too. Almost", *NASA*, retrieved from: https://www.nasa.gov/vision/universe/roboticexplorers/robots_like_people.html (accessed May 01, 2019).
8. "The Perfect Team: People and Robots, The Economist – Films" *The Economist*, https://films.economist.com/winningedge/the_perfect_team_people_and_robots?linkId=100000005440691 (accessed May 01, 2019).
9. Walsh, N.P. (September 02, 2001). "Alter Our DNA or Robots will take over, warns Hawking", *The Guardian*, retrieved from: https://www.theguardian.com/uk/2001/sep/02/medicalscience.genetics (accessed May 31, 2019).
10. Don Tapscott, a Canadian author, consultant and Executive Chairman, Blockchain Research Institute. The same view has been articulated by Paul Viga and Michael Carey in their 2018 book titled, *The Truth Machine: The Blockchain and Future of Everything.*
11. Narayanan, et.al. (February 9, 2016), *Bitcoin and Crypto currency Technologies*, retrieved from: https://www.lopp.net/pdf/princeton_bitcoin_book.pdf (accessed May 31, 2019).
12. Hassell, J. (April 14, 2016). "What is Blockchain and how does it work?", *CIO*, https://www.cio.com/article/3055847/what-is-blockchain-and-how-does-it-work.html (accessed May 31, 2019).
13. Hassel, J., Ibid
14. Siddiqui, I. (April 25, 2018). "What the hell is Blockchain and How does it Works?", *Medium*, retrieved from https://medium.com/coinmonks/what-the-hell-is-blockchain-and-how-does-it-works-simplified (accessed June 5, 2019).
15. Vigna, P. & Michael, C. (2018). *The Truth Machine: The Blockchain and the Future of Everything,* St. Martin's Press.
16. Safra Catz is Chief Executive of Oracle
17. Dale, S. (November 20, 2017). "Automation and AI – What does the future of work look like?" *The Future of Work*, retrieved from: http://www.stephendale.com/2017/11/20/automation-and-ai-what-does-the-future-of-work-look-like/ (accessed 1 May, 2019).
18. "Benefits of Automation", *Productivity Inc.*, retrieved from: https://www.productivity.com/benefits-of-automation/. Accessed 1 May, 2019.
19. Anthony J Stieber is author of the book *Breaking into Information Security Crafting a Custom Career Path to Get the Job You Really Want* (December 2015).

20. Stuart Chase (1888-1985) was an American Economist, Social Theorist, and Writer. The quote is from www.goodreads.com , taken most probably from his book: "The Most Probable World" (1981).
21. "AM Basics - Additive Manufacturing (AM)", *Additive Manufacturing*, retrieved from: http://additivemanufacturing.com/basics/ (accessed May 01, 2019).
22. "3D Printing", *Lexico Dictionary*, retrieved from: https://en.oxforddictionaries.com/definition/us/3d_printing (accessed May 01, 2019).
23. AM Basics, *op. cit.*
24. Professor Neil Jacobstein chairs the Artificial Intelligence and Robotics Tract at Singularity University on the NASA Research Park Campus
25. Said Brendan O'Brien, Co-founder of Aria Systems.
26. "Internet of things (IoT)" (March 13, 2019). *IoT Agenda*, retrieved from: https://internetofthingsagenda.techtarget.com/definition/Internet-of-Things-IoT (accessed May 01, 2019).
27. Porters, M. and Heppelman, J. (November 2014), "How Smart, Connected Products Are Transforming Competition", *Harvard Business Review*, retrieved from: https://hbr.org/2014/11/how-smart-connected-products-are-transforming-competition (accessed May 01, 2019).
28. Grant Notman, Head of Sales and Marketing at Wood & Douglas
29. "8 industries where IoT is working the best", *Network World*, retrieved from: https://www.networkworld.com/article/3200783/8-industries-where-iot-is-working-the-best (accessed May 01, 2019).
30. Mehul Nayak, Digital Marketing Consultant at IoT Solutions provider company in Chicago, US.
31. "Multi-trillion Dollar Market", Report by PricewaterhouseCoopers (2014).
32. In a T.V. Interview (November 26, 2018)
33. "Ultimate Guide to Understand Augmented Reality (AR) Technology", *Reality*, retrieved from: https://www.realitytechnologies.com/augmented-reality/ (accessed May 01, 2019).
34. Ibid.
35. Muchaneta Kapfunde, Fashion Tech Consultant, Founder and Editor-in-Chief of FashNerd.com

Chapter 5

1. Harai , Y.L. (2018). *21 Lessons for the 21st Century,* Israel: Spiegel and Grau, Jonathan Cape.
2. "McKinsey Global Institute Report" (September 2018), reviewed by Irving Wlasawsky-Berger for *Wall Street Journal* (November 16, 2018).

3. www.statista.com
4. McKinsey Report, *op. cit.*
5. Ibid.
6. Similar results are also available in PricewaterhouseCooper (PwC) Report: "Sizing the Prize" (2018), PwC's Global Artificial Intelligence Study.
7. The concept of "S-curve" was put forth by Professor Eric Bryn Jolfsson, of MIT in November 2017.
8. Irving Wladawsky - Berger, *op. cit.*
9. Jayne, T.S. et.al. (2018), "Future of work in African Agriculture", *ILO*, Working Paper 25.
10. "The Impact of Technology on the Quality and Quantity of Jobs" (January 6, 2018), *ILO*, retrieved from: https://www.ilo.org/wcmsp5/groups/public/---dgreports/---cabinet/documents/publication/wcms_618168.pdf. (accessed May 07, 2019).
11. Ibid.
12. Young, J. (January 21, 2019). "Global e-commerce Sale Grow 18% in 2018", *Digital Commerce 360*, retrieved from: https://www.digitalcommerce360.com/article/global-ecommerce-sales/. (accessed May 01, 2019).
13. 2018 Study by Jacob Poushter, Caldwell Bishop and Hanyu Chwe, Pew Research Centre.
14. Zaidi, D. (April 22, 2018). "The 3 Most Valuable Applications of AI in Health care", *Venture Beat*, Retrieved from https://venturebeat.com/2018/04/22/the-3-most-valuable-applications-of-ai-in-health-care/. (accessed May 07, 2019).
15. "Applying AI for Social Good", *McKinsey Global Institute* (November 2018).
16. Ibid.
17. Ibid.
18. McKinsey (2018), *op. cit.*
19. Rainforest Connection is a non-profit technology start up-a digital social connection, that develops and implements free, open source software and systems to help preserve rainforests.
20. McKinsey, *op. cit.*
21. Souto, E. & Davepon, C. (November 13, 2017). "Technological Unemployment: Why Keynes Is More Relevant Than Ever", *Medium*, retrieved from: https://medium.com/new-tech-revolution-sciencespo/technological-unemployment-why-this-keynesianism-term-is-more-than-ever-up-to-date-479a31d5592d. (accessed May 06, 2019).
22. Keynes, J.M. (1930). "Economic Possibilities for Our Grandchildren". An essay originally published in 1930, reproduced in *Essays in Persuasion* (2010) Palgrave Macmillan, London.

23. Mashelkar, R.A., "The Future of Technology and Jobs", an Interview by Arun Kumar Tripathi, Ubiquity (April 2016).

24. Souto, E. (2017), *op. cit.*

25. McAfee & Brynjolfsson, E. (2011), "Race against the Machine: How the Digital Revolution Is Accelerating Innovation, Driving Productivity, and Irreversibly Transforming Employment and the Economy" *Digital Frontier Press.*

26. Frey, C.B. and Osborne, M. (September 17, 2013). "The Future of Employment: How Susceptible Are Jobs to Computerisation", Retrieved from https://www.oxfordmartin.ox.ac.uk/downloads/academic/The_Future_of_Employment.pdf, (accessed May 9, 2019).

27. Frey, C.B. et.al. (January 2016). "Technology at Work v2.0", *Citi GPS: Global Perspectives & Solutions*, retrieved from: https://www.oxfordmartin.ox.ac.uk/downloads/reports/Citi_GPS_Technology_Work_2.pdf, (accessed May 9, 2019).

28. Osborne, M. and Frey, C.B. (April 13, 2018). "Automation and the Future of Work – Understanding the Numbers", *Oxford Martin School, University of Oxford*, Retrieved from https://www.oxfordmartin.ox.ac.uk/blog/automation-and-the-future-of-work-understanding-the-numbers/. (accessed May 09, 2019).

29. Ibid.

30. Mashelkar (2016), *op. cit.*

31. Oxford Martin School Report (2013), *op. cit.*

32. Akkavi, Y. (April 2, 2018) "Worried About a Robot Taking Your Job? Actually, it will be Your Next Boss", *Inc.*, retrieved from: https://www.inc.com/yazin-akkawi/the-ceo-of-future-is-literally-a-machine.html. (accessed May 07, 2019).

33. Cookson, C. (February 14, 2016). "AI and Robots Threaten to Unleash Mass Unemployment", *Financial Times*, retrieved from: https://www.ft.com/content/063c1176-d29a-11e5-969e-9d801cf5e15b (accessed May 07, 2019).

34. Novosilska, L. (November 30, 2018) "5 Ways Artificial Intelligence is Impacting the Automotive Industry", Ignite, retrieved from: https://igniteoutsourcing.com/automotive/artificial-intelligence-in-automotive-industry/. (accessed 7 May, 2019).

35. Mashelkar in his interview, *op. cit.* gives this example.

36. Arntz, M., Gregory, T. and Zierahn, U. (2016) "The Risk of Automation for Jobs in OECD Countries: A Comparative Analysis," *OECD Social, Employment and Migration Working Paper No. 189*, retrieved from: http://www.ifuturo.org/sites/default/files/docs/automation.pdf. (accessed 7 May, 2019).

37. Pethokoukis, J. (June 6, 2016). "What the story of ATMs and bank tellers reveals about the 'rise of the robots and Jobs", *AEI*, retrieved from: http://www.aei.org/publication/what-atms-bank-tellers-rise-robots-and-jobs/. (accessed May 07, 2019).
38. Ibid.

Chapter 6

1. Kuznets, S. (1955), "Economic Growth and Income Inequality", *The American Economic Review*, 45(1), pp 1-25
2. Timbergen J., (1970), "A Positive and Normative Theory of Income Distribution", *Review of income and Wealth*, 16(1), pp. 221-234
3. Thomas Edsall, (January 28, 2018) "Capitalism v. Democracy", *The New York Times*, retrieved from: https://www.nytimes.com/2014/01/29/opinion/capitalism-vs-democracy.html. (accessed May 18, 2019)
4. Solow, R.M. (April 22, 2014). "Thomas Piketty is Right", *The New Republic.*
5. Piketty, T. (2014). "Capital in the Twenty-first Century" (Translated into English by A. Goldhammer), Cambridge, MA: The Belknap Press of Harvard University Press.
6. Milanovic, B. (2016). "Global Inequlity: A New Approach for the Age of Globalization", *Harvard University Press*, as quoted in Bauer Johannes, M. *Internet and Income Inequality: Socio-economic Challenges in a Hyperconnected Society*, SSRN Report on "Telecommunications policy" (July 25, 2017), Downloaded on May 18, 2019.
7. Bourguignon, F. and Morrison, C. (2002). "Inequality Among World Citizens: 1920 – 1992", *American Economic Review*, 92(4), pp. 727-744.
8. As quoted in Johannes Bauer Report (2017), *op. cit.*
9. Milanovic (2012), Ibid.
10. Ibid.
11. "In it Together: Why Less Inequality Benefits All" (2015), OECD Paris, p.35.
12. Ibid.
13. A "Digital Divide" refers to any uneven distribution in the access to and use of digital technology between any number of distinct groups, within or across countries.
14. See Bauer, J.M. (2017), *op. cit.*; Bourguignon and Morrison (2002), *op. cit.* Also see Bourguignon (2015). "The Globalization of Inequality", *Princeton University Press*, as well as Krugman, P. (1979), "A Model of Innovation, Technology Transfer, and World Distribution of Income", *Journal of Political Economy*, 87 (2), pp. 253-266.

15. For example, Bauer (2017), *op. cit.*
16. Frankfirt, H.G (2015). "On Inequality", *Princeton University Press.*
17. Wertime. K., "Closing the Digital Divide in India", *WPP*, retrieved from: https://www.wpp.com/featured/diversity-and-inclusion/closing-the-digital-divide-in-india. (accessed May 24, 2019).
18. Garimella. K., (August 7, 2018). "Job Loss from AI? There's more to Fear!", *Forbes*, retrieved from: https://www.forbes.com/sites/cognitiveworld/2018/08/07/job-loss-from-ai-theres-more-to-fear/. (accessed June 08, 2019).
19. Ibid.

Chapter 7

1. Lagarde, C. (June 8, 2019). "The Next Steps for International Cooperation in Fintech" Opening Remarks at G-20 High level Seminar "*Our Future in the Digital Age*", Fukuoka, Japan, retrieved from: https://www.imf.org/en/News/Articles/2019/06/07/sp060819-lagarde-the-next-steps-for-international-cooperation-in-fintech. (accessed June 23, 2019
2. Jadhav, N. (2006). "Monetary Policy, Financial Stability and Central Banking in India", New Delhi: Macmillan India Ltd, p. 166.
3. Sheng, A. (May 26, 2018). "From Industrial 4.0 to Finance 4.0", *The Star Online*, retrieved from: https://www.thestar.com.my/business/business-news/2018/05/26/from-industrial-40-to-finance-40/. (accessed June 23, 2019). Andrew Sheng was Deputy Chief Executive of Hong Kong. Monetary Authority and later, Chairman, Hong Kong Securities and Futures Commission.
4. Ibid.
5. Lagarde, C. (June 8, 2019). "IMF's Christine Lagarde Warns of Disruption from Fintechs", *DW*, Retrieved from https://www.dw.com/en/imfs-christine-lagarde-warns-of-disruption-from-fintechs/a-49110435. (accessed June 23, 2019).
6. King, B. (2019). *Bank 4.0: Banking Everywhere, Never at a Bank*, Wiley. His earlier books include *Bank 2.0: How Customer Behaviour and Technology will Change the Future of Financial Services* (2010) and *Bank 3.0: Why Banking is No Longer Somewhere You Go But Something You Do* (2012).
7. Mladenovic, S. (September 28, 2018). "Banking in Industry 4.0: Robotic Automation as an Answer to the Challenge of Tomorrow", *Comtrade System Integration*, retrieved from: https://www.comtradeintegration.com/en/banking-in-industry-4-0-robotic-automation-as-an-answer-to-the-challenges-of-tomorrow/. (accessed June 23, 2019).
8. King, B., *op. cit.*
9. King, B., Ibid.
10. King, B., Ibid.

11. King, B., Ibid.
12. Sheng, A., *op. cit.*
13. Sheng, A., Ibid.
14. Sheng, A., Ibid.
15. Singh, R. (January 22, 2019). "Finance 4.0: The Race to Rebalance the World", *Financial IT*, Retrieved from https://financialit.net/blog/financial-services/finance-40-race-rebalance-world. (accessed June 23, 2019).
16. Sheng, A., *op. cit.*
17. Sheng, A., Ibid.
18. Marous, J. (August 27, 2018), "The Future of Banking: Fintech or Techfin?", *Forbes*, retrieved from: https://www.forbes.com/sites/jimmarous/2018/08/27/future-of-banking-fintech-or-techfin-technology/#b11a7c55f2d5. (accessed June 23, 2019).
19. Jadhav, N., *op. cit.*, p.53.
20. Marous, J., *op. cit.*
21. Marous, J., Ibid.
22. Cook, J. (January 23, 2015). "Jack Ma: Here's How Alibaba Will Become Bigger Than Wall Mart", *Business Insider India*, retrieved from: https://www.businessinsider.in/JACK-MA-Heres-How-Alibaba-Will-Become-Bigger-Than-Walmart/articleshow/45993028.cms. (accessed June 24, 2019).
23. King, B., *op. cit.*, p.189.
24. Ernst & Young (2019). "Global FinTech Adoption Index", retrieved from: https://assets.ey.com/content/dam/ey-sites/ey-com/en_gl/topics/banking-and-capital-markets/ey-global-fintech-adoption-index.pdf. (accessed June 24, 2019).
25. King, B., 2019, *op. cit.*, p. 189
26. King, B., 2019, *op. cit.*, p. 194.
27. "The Future of FinTech and Banking: Digitally Disrupted or Re-imagined?" (2015), *Accenture*, retrieved from: https://www.accenture.com/_acnmedia/Accenture/Conversion-Assets/DotCom/Documents/Global/PDF/Dualpub_11/Accenture-Future-Fintech-Banking.pdf (accessed June 24, 2019).
28. Drummer, D. et.al. (2016). "FinTech - Challenges and Opportunities", *McKinsey & Company.*
29. "Financial Services Technology 2020 and Beyond: Embracing Disruption" (2016), *PwC*, retrieved from: https://www.pwc.com/gx/en/financialservices/assets/pdf/technology2020-and-beyond.pdf. (accessed June 24, 2019).
30. King, B., *op. cit.*, p.201.

31. Ibid. at p. 204.
32. Young, J. (January 21, 2019). "Global e-Commerce Sales Grow 18% in 2018", *Digital Commerce 360*, retrieved from: https://www.digitalcommerce360.com/article/global-ecommerce-sales/, (accessed June 24, 2019)
33. Ibid.
34. Media Centre, *Bain & Company*, (November 15, 2018) https://www.bain.com/about/media-center/press-releases/2018/fall-luxury-goods-market-study/
35. U.S Census Bureau News (May 17, 2019), Quarterly Retail E-commerce Sales, *U.S. Department of Commerce*, retrieved from: https://www.census.gov/retail/mrts/www/data/pdf/ec_current.pdf(accessed June 24, 2019)
36. Young, J., *op. cit.*
37. Ibid.
38. "Technology Trends in E-Commerce" (June 05 2018), *Netscribes*, retrieved from: https://www.netscribes.com/ecommerce-technology-trends/. (accessed June 24, 2019).
39. Ibid.
40. "Gartner Says By 2018, 50 Percent of Consumers in Mature Markets Will Use Smartphones or Wearables for Mobile Payments" (December 15, 2015), *Gartner*, retrieved from: https://www.gartner.com/en/newsroom/press-releases/2015-12-15. (accessed June 24, 2019)
41. "Forrester Data: Mobile Payments Forecast, 2016 To 2021 (US), Q4 2016 Update", (February 6 2017), *Forrester*, retrieved from: https://www.forrester.com/report/Forrester+Data+Mobile+Payments+Forecast+2016+To+2021+US+Q4+2016+Update/-/E-RES137043. (accessed June 24, 2019)
42. Technology Trends in E-Commerce, *op. cit.*
43. The FATF is an intergovernmental organization founded in 1989 on the initiative of the G-7 countries to develop policies to combat money laundering. From 2001, its mandate also covers terrorism financing.
44. "Virtual Currencies: Key Definitions and Potential AML/CFT Risks" (2014*), Finanical Action Task Force*, retrieved from: https://www.fatfgafi.org/media/fatf/documents/reports/Virtual-currency-key-definitions-and-potential-aml-cft-risks.pdf, (accessed June 24, 2019)
45. Bank of International Settlements (2018), *Annual Economic Report*, retrieved from: https://www.bis.org/publ/arpdf/ar2018e.pdf. (accessed June 24, 2019).
46. Ibid.
47. Memoria, F. (December 2, 2017). "Nobel Laureate Economist Predicts a Bitcoin Crash "Won't Go to Zero, It Will Just Come Down", *CCN*, retrieved from: https://www.ccn.com/nobel-laureate-economist-

predicts-bitcoin-crash-wont-go-to-zero-it-will-just-come-down, (accessed June 24, 2019).

48. Kelly, B. (2015). "The Bitcoin Big Bank: How Alternative Currencies Are About to Change the World", *Wiley*.
49. Wilson, C. (2019). "Cryptocurrencies: The Future of Finance?" *Contemporary Issues in International Political Economy*, 359–394.
50. Ibid.

Chapter 8

1. Spokesperson for American Civil Liberties Union (ACLU) as quoted by Kirk Baird in *The Blade* (July 9, 2017).
2. From the interview conducted during Annual Clinton Global Initiative Meeting (September 23, 2013).
3. Mill, J.S. (1859). *On Liberty*, Batoche Books, p. 6 &13.
4. *Harvard Law Review* (1890), Vol. 4, p. 193.
5. The technology that they were talking about was printing technology.
6. Miller, Arthur R. "The Assault on Privacy: Computers, Data Banks and Dossiers" (1971), *University of Michigan Press*.
7. Thomson, J.J. (1975). "The Right to Privacy", *Philosophy and Public Affairs*.
8. Inness, Julie C. (1992). "Privacy, Intimacy and Isolation", *Oxford University Press*.
9. Professor Daniel Solove in his definitive work on Privacy, titled: "Understanding Privacy", *Harvard University Press* (2008), p. 12-13.
10. Ibid.
11. Godkin, E.L. (1880), "Libel and its Legal Remedy", *Atlantic Monthly*, pp 729-39
12. Solove (2008), *op. cit.*
13. Solove (2008), *op. cit.* and Posner, R.A. (1983). "The Economics of Justice", *Harvard University Press*, p. 271.
14. Solove (2008), *op. cit.*
15. Altman, I. (1975). *The Environment and Social Behaviour: Privacy, Personal self, Territory, and Crowding,* Monterey: Brooks/Cole Publishing Company.
16. Kupfer, J., "Privacy, Autonomy and Self-Concept", *American Philosophical Quarterly,* Vol. 24, No. 1 (January, 1987), pp. 81-89
17. Rachels, J. (1975). "Why Privacy is Important", *Philosophy & Public Affairs*, 4(4) pp. 323-33.
18. Growman, M., (January 15, 2019). "As Technology Advances, What Will Happen With Online Privacy?" *Forbes*, retrieved from: https://www.forbes.com/sites/quora/2019/01/15/as-technology-advances-what-will-happen-with-online-privacy/#61352e241c45. (accessed May 10, 2019).

19. Ibid.
20. "Five Ways Your Privacy is Compromised without You Realising" (September 28, 2018). *Privacy International*, Retrieved From: https://privacyinternational.org/feature/2319/five-ways-your-privacy-compromised-without-you-realising. (accessed May 14, 2019).
21. Berman, J. & Mulligan, D. (1999) "Privacy in the Digital Age: Work in Progress", *Nova Law Review*, 23, 2.
22. Calo R.M, "Robots and Privacy", in *Robot Ethics: The Ethical and Social Implications of Robotics* (2011), eds. Bekey, P., Abney, G., Lin, K., Boston: MIT Press.
23. Francisco J.R.L., et.al, (2017). "Cybersecurity of Robotics and Autonomous Systems: Privacy and Safety", Robotics: Legal, Ethical and Socio Economic Impact, *Intech Open.*
24. Calo, RM, *op. cit.*
25. Francisco, et al, *op. cit.*
26. Van den Hoven van Genderen, R., (2017). "Privacy and Data Protection in the Age of Pervasive Technologies in AI and Robotics", *European Data Protection Law Review*, 3. 338-352.
27. Dag Svrdal, et.al. (January 2007). "He Knows When You Are Sleeping – Privacy and the Personal Robot Companion", *AAAI*, p. 28
28. Ibid.
29. Calo, R.M., *op. cit.*
30. Ibid.
31. Dag Sverre Syrdal, *op. cit.*
32. Ibid.
33. Ibid.
34. Mitchell, G. (n.d.). "How Much Data is on the Internet?" *Science Focus*, retrieved from: https://www.sciencefocus.com/future-technology/how-much-data-is-on-the-internet/. (accessed May 14, 2019).
35. Parker, C.B. (January 12, 2015). "Stanford Research Finds that Computers Are Better Judges of Personality than Friends and Family", *Stanford*, retrieved from: https://engineering.stanford.edu/magazine/article/stanford-research-finds-computers-are-better-judges-personality-friends-and-family. (accessed May 14, 2019).
36. iNews Article, *op. cit.* The Issue of Surveillance is discussed, in detail, in Chapter 12.
37. Solove (2008), *op. cit.*
38. Ibid.
39. Ibid.
40. Quinn, Michael J. (2009). "Ethics for the Information Age", *Pearson.* The book has gone into several editions.

41. Jensen, C. and Potts, C. (2004). "Privacy Policies as Decision-making Tools: An Evaluation of Online Privacy Notices", *CHI '04 Proceedings of the SIGCHI Conference on Human Factors in Computing Systems*, p. 471-478.
42. Bhatia, R. (January 30, 2018). "Is Deep Learning Going to be Illegal in Europe?", *Analytics India*, Retrieved From: https://www.analyticsindiamag.com/deep-learning-going-illegal-europe/. (accessed May 14, 2019).
43. The concept of Privacy is embodied implicitly within the First, Third, Fourth, Fifth and Fourteenth Amendments.
44. Bruggemeier, G. (2009). "Protection of Personality Interests in Continental Europe: The Examples of France, Germany, Italy, and a European Perspective", Rights of Personality in Scots Law: A Comparative Perspective, *Dundee University Press.*
45. Emejuru, C.T. (2015). "Wainwright v. United Kingdom: Filling the Gaps of Common Law Rights to Privacy Under The European Convention On Human Rights", *Journal of Law, Policy and Globalization*, 35. 186-196.
46. Palmer, V.V. (2010). "Three Milestones in the History of Privacy in the United States", *Tulane European & Civil Law Forum*, 26, p.67
47. Ibid.
48. "Big Brother Watch & others. v. United Kingdom", *European Court of Human Rights* (2018)
49. Michael, P.L. (May 2015). "The Philosophy of Privacy", *The Guardian,* retrieved from: https://www.theguardian.com/technology/2015/may/07/surveillance-privacy-philosophy-data-internet-things. (accessed December 01, 2018).
50. Justice K.S. Puttaswamy (Retd.) v. Union of India, (2017) 10 SCC 1.

Chapter 9

1. Roxane Gay is an American writer, professor, editor and commentator. She is the author of the New York Times bestselling essay collection, *Bad Feminist* (2014)
2. Both quotes are from www.brainquote.com. (accessed June 20, 2019).
3. Obar, J.A. and Wildman, S. (2015). "Social Media Definition and the Governance Challenge", *Telecommunications Policy*, 39 (9), pp. 745-50,
4. "Number of Social Media Users Worldwide, 2010-2021", *Statista*, retrieved from: www.statista.com. (accessed May 22, 2019).
5. Statista, *op. cit.*
6. Karnik, K. (2018), *op. cit.*, pp. 93-94.
7. Ibid, p.94.
8. Amedie, J. (2015), "The Impact of Social Media on Society," *Santa Clara University*, Retrieved From: http://scholarcommons.scu.edu/engl- 176/2. (accessed May 22, 2019).

9. Leaver, T. (2013), "The Social Media Contradiction: Data Mining and Digital Death", *M/C Journal,* 16(2)
10. Keen, A. (2007). *The Cult of the Amateur,* Penguin Random House, p.15
11. Karnik, K. (2018), *op. cit.*
12. Newman, N. and Levy, D. (2013). "Digital News Report", *Return's Institute.*
13. Amedie, J., (2015). "The Impact of Social Media on Society", *Santa Clara University*, 176/2.
14. Weimann, G. (2008). "The Psychology of Mass-Mediated Terrorism", *American Behavioral Scientist*, Vol.52, pp 69-86
15. Brown, I. (2015). "Social Media Surveillance", *The International Encyclopedia of Digital Communication and Society*, 1-7.
16. Ball, K., Lyon, D. & Haggerty, K.D. (Eds.) (2018). *Routledge Handbook of Surveillance Studies*, Routledge, 123.
17. Prasanth, R. (June 8, 2018). "Social Media Hub or Tool of Mass Surveillance", *News Click*, retrieved from: https://www.newsclick.in/social-media-hub-or-tool-mass-surveillance, (accessed June 20, 2018.
18. Albrechtslund, A. (2017). "Online Social Networking as Participatory Surveillance", *First Monday*, 13 (3).
19. Ibid.
20. Christopher Wylie reportedly had been on anonymous source for the 2017 article in the observer, titled, "The Great British Brexit Robbery".
21. Cohen, J. (2016). "What Privacy is For?", *Harvard Law Review*, 126, pp. 1909-10
22. Ibid.
23. Burgess M. (January 21, 2019). "What is GDPR? The Summary Guide to GDPR compliances in the UK", *Wired,* retrieved from: http://www.wired.co.uk/article/what-is-gdpr-uk-eu-legislation-compliance-summary-fines-2018. (accessed June 20, 2019).

Chapter 10

1. Comey, J. (Ex FBI Director).
2. International Telecommunication Union (ITU), retrieved from: https//www.itu.int/ITU-D/Documents. (accessed May 2019).
3. Ibid.
4. Issued by the World Economic Forum, 14th Edition, 7.
5. GPRS is General Packet Radio Service which allows transfer of data over cellular networks.
6. Posner, R.A. (2008). "Privacy, Surveillance and Law", *The University of Chicago Law Review*, Vol.75, pp.251.

7. Kan, M. (October 26, 2016). US Correspondent of Computer World, retrieved from: computerworld.com. (accessed May 2019).
8. Global Risk Report, 2019, *op. cit.,* p.7
9. National Defence College of India (1996).
10. Clemente, J.F. (September 2018). "Cyber-Security for Critical Energy Infrastructure", *Naval Postgraduate School, California*, retrieved from: https://calhoun.nps.edu/.
11. Slay, J. & Miller, M. "Lessons Learned from the Maroochy Water Breach: Critical Infrastructure Protection", *International Federation for Information Processing*, Vol. 253, pp. 73-82.
12. "Symantec Internet Security Threat Report" (February 2003), *Symantec*. (which provides a six-month update of Internet threat activity).
13. C. Kenneth, "Ensuring and Insuring Critical Information Infrastructure Protection", Report of the 2005, Rueschlikon Conference on Information Policy, Rueschlikon, Switzerland.
14. Bossert, T.P. (December 18, 2017). "It's official: North Korea is Behind Wanna Cry," *The Wall Street Journal,* (accessed June 14, 2019.
15. Limaye, V. (October 21, 2018). "Here's a Cyber Security Roadmap for the Financial Sector", *Rediff*, retrieved from: https://www.rediff.com/business/column/. (accessed June 14, 2019).
16. Symantec, *op. cit.*
17. Ibid.
18. Saravade, N. and Bhalla, A. "Emerging Trends and Challenges in Cybersecurity", *White Paper issued by Reserve Bank Information Technology Pvt Ltd* (REBIT), retrieved from: https://rebit.org.in/whitepaper/emerging-trends-and-challenges-cyber-security.
19. "Global Economic Crime and Fraud Survey 2018", *PricewaterhouseCoopers (PwC),* retrieved from: www.pwc.com.
20. Ibid.
21. Cybercrime, Organized Crime, and Societal Responses – Springer.
22. Retrieved from: https://link.springer.com/content/pdf/10.1007%2F978-3-319-44501-4.pdf. (accessed July 03, 2019).
23. "Social engineering fraud – Interpol", retrieved from: https://www.interpol.int/en/Crime-areas/Financial-crimes/Social-engineering-fraud/Types-of-social-engineering-fraud. Accessed 3 Jul. 2019.
24. "Cyber theft and the loss of intellectual property" (July 25, 2016), *Deloitte Review,* retrieved from: https://www2.deloitte.com/insights/us/en/deloitte-review/issue-19/loss-of-intellectual-property-ip-breach.html. (accessed July 03, 2019).
25. Vegh, S. (2002). "Hacktivists or Cyberterrorists? The Changing Media Discourse on Hacking", *First Monday,* retrieved from: https://firstmonday.org/article/view/998/919. (accessed July 03, 2019).

26. "Cyber Warfare", *RAND Corporation*, retrieved from: https://www.rand.org/topics/cyber-warfare.html. (accessed July 03, 2019).
27. Global Security Index (GCI) of the UN.

Chapter 11

1. Environmentalist and Vice President of the US (1993-2001). The quote is from his book: *The Future: Six Drivers of Change* (2013), Random House, New York (p. 86).
2. Rainer, E, "The Role of the State Security Service (Stasi) in the context of international clinical trials conducted by western pharmaceutical companies in Eastern Germany", *Plus One Journal*, retrieved from: https://journals.plos.org/plosone/article.
3. Lyon, D. (2007). "Surveillance Studies: An Overview", *Cambridge: Polity Press.*
4. Chayco, M. (2016), "Superconnected: The Internet, Digital Media and Techno-social Life", *Sage Publications.*
5. Ibid.
6. Ackerman, S. and Theilman, S. (February 9, 2016). "US intelligence chief: we might use the internet of things to spy on you", *The Guardian*, retrieved from: https://www.theguardian.com/technology/2016/feb/09/internet-of-things-smart-home-devices-government-surveillance-james-clapper. (accessed July 05, 2019).
7. On March 3, 1991, Rodney King, an African-American construction worker was violently beaten by the Los Angeles Police Department (LAPD). A civilian filmed the incident of police brutality from his nearby balcony and sent the footage to local news station. The incident was covered by news media around the world.
8. Figgis, J.N. (1922). *The Divine Right of Kings*, University Press. Also, Fischlin, D. & Fortier, M. (Eds.). (1996). *The True Law of Free Monarchies,* Centre for Reformation and Renaissance Studies.
9. Jenks, E. (1900). "A History of Politics", *Biblio Life* (2009).
10. Locke, J. (1689). *Second Treatise of Government,* London: Awnsham Charchill.
11. Justice K.S. Puttaswamy v. Union of India, (2017) 10 SCC 1.
12. Chesterman, S. (2011). "One Nation Under Surveillance: A New Social Contract to Defend Freedom Without Sacrificing Liberty", *Oxford University Press.*
13. Lee, T.B. (n.d.). "Principles for a Contract for the Web", *A Contract for the Web,* retrieved from: https://contractfortheweb.org/. (accessed May 23, 2019).

14. Odell, M. in *Financial Times* (August 1, 2005).
15. Reported by Sanger, D. in *New York Times* (December 26, 2014).
16. Reported by Gellman, B. in *Washington Post* (December 4, 2013).
17. Ibid.
18. Reported by *BBC News* (April 13, 2018).
19. Klein, N. (May 29, 2008). "China's All-Seeing Eye", *Rolling Stone.*
20. As reported in *Wired Magazine* (July 18, 2007).
21. As reported in *The Guardian* (June 6, 2013).
22. Marks, P. (June 9, 2006)."Pentagon Sets its Sights on Social Networking Website", *New Scientist.*
23. Foucault, M. (1995), *Discipline and Punish*, New York: Vintage Books, pp. 201-02
24. Ibid.
25. "Is the US Turning in to a Surveillance Society?" (March 13, 2009), *American Civil Liberties Union.*
26. Ibid.
27. Houstan, T. (April 29, 2017). "Mass Surveillance and Terrorism: Does PRISM Keep Americans Safer?" *TRACE*, retrieved from: http://trace.tennessee.edu/cgi/viewcontent. (accessed May 23, 2019).
28. Browne, S. (October 2015). "Dark Matters: On the Surveillance of Blackness", *Duke University Press*, p. 224.

Chapter 12

1. Yudkowsky, E. (2008), "Artificial Intelligence as a Positive and Negative Factor in Global Risk" In *Global Catastrophic Risks*, eds. Nick Bostrom and Milan M. Ćirković, Oxford University Press, pp. 308–345.
2. Walter Lippman (1889-1974) was an American Writer, Reporter and Political Commentator who apparently made this statement in 1943 i.e, towards the end of the World War II, in his book: *US Foreign Policy: Shield of the Republic,* Little, Brown and Company.
3. Brown, H. (1983). *Thinking About National Security: Defence and Foreign Policy in a Dengerous World,* Westview Press. Harold Brown was the US Secretary of Defence during the period, 1977-1981, under President Jimmy Carter.
4. "Economic Security for a Better World" (2004), *International Labour Organization (ILO).*
5. Ibid.
6. United States v. Robel, 389 US 258 (1968).
7. Ralph Nader quote is from nader.org.

8. Verbruggen, M. and Boulanin V. (November 2017). "Mapping the Development of Autonomy in Weapon Systems", *Stockholm International Peace Research Institute*, retrieved from: https://www.sipri.org/publications/2017. (accessed April 08, 2019).
9. *"Royal Navy Looking to Use Artificial Intelligence"* (October 17, 2016), *Roke, UK* retrieved from: https://www.roke.co.uk/media/news/. (accessed April 08, 2019).
10. "Innovation Challenge - Artificial Intelligence in Royal Navy Warships" (October 17, 2016) *Tech UK*, retrieved from: . (accessed April 08, 2019).
11. Welsh, S. (June 5, 2015). "Lethal Autonomy in Autonomous Unmanned Vehicles", *Center for International Maritime Security*, retrieved from: http://cimsec.org/lethal-autonomy-autonomous-unmanned-vehicles/16732. (accessed May 9, 2019).
12. Maaike, V. and Boulanin, V. (November 2017). "Mapping the Development of Autonomy in Weapon Systems", *Stockholm International Peace Research Institute*, retrieved from: https://www.sipri.org/publications/2017/other-publications/mapping-development-autonomy-weapon-systems. (accessed May 9, 2019).
13. Nair, G.A. (October 11, 2017). "Rolls-Royce to 'Man' Autonomous Ship with Google AI software", *The Hindu Business Line*, retrieved from: https://www.thehindubusinessline.com/economy/logistics. (accessed May 9, 2019).
14. Mukherjee, T. (July 16, 2018). "Securing the Maritime Commons: The Role of Artificial Intelligence in Naval Operations, *Observer Research Foundation (ORF) Occasional Paper*, retrieved from: https://www.orfonline.org/research/42497-a-i-in-naval-operations-exploring-possibilities-debating-ethics/#_ednref33
15. Reddy R.S. (February 06, 2017). "The Swarm of War: Is India Ready?", *The Wire*, retrieved from: https://thewire.in/external-affairs/swarm-war-india-ready. (accessed May 9, 2019).
16. Lachow, I., (February 28, 2016). The Upside and Downside of Swarming Drones, *Bulletin of Atomic Scientists*, 73:2, pp. 96-101, retrieved from: https://www.tandfonline.com/doi/full/10.1080/00963402.2017.1290879. (accessed May 9, 2019).
17. Bendett, S. (January 10, 2017). "How Russia's Military Plans to Counter the Pentagon's Drone Swarms", The National Interest, retrieved from: https://nationalinterest.org/blog/the-buzz/how-russias-military-plans-counter-the-pentagons-drone-19014.
18. Popescu, A. (June 6, 2016). "Cody wilson: The Man who Wants Americans to Print Their Own 3-D Guns", *The Guardian*, retrieved from: https://www.theguardian.com/us-news. (accessed May 9, 2019).
19. Retrieved From: https://www.wired.com. (accessed May 9, 2019).

20. Tirone, D.C. & Giley, J. (July 19, 2016). "3D Printing: New Threat to gun control and security policy?", *The Conversation*, retrieved from: https://theconversation.com/3d-printing-a-new-threat-to-gun-control-and-security-policy-61416. (accessed May 9, 2019).
21. Sebastian, A. (July 26, 2012). "The World's First 3D Printed Gun", *Extreme Tech*, retrieved from: http://www.extremetech.com/extreme/133514-the-worlds-first-3d-printed-gun, (accessed May 9 2019)

    Greenberg, A. (August 23, 2012). "Wiki Weapon Project Aims to create a gun anyone can 3D print at home", *Forbes*, retrieved from https://www.forbes.com/sites/andygreenberg/2012/08/23/wiki-weapon-project-aims-to-create-a-gun-anyone-can-3d-print-at-home/#4fc356b43860. (accessed May 9 2019).
22. "U.S. Navy Missile Program Uses First 3D Printed Component" (March 18, 2016), *Reuters,* retrieved from: http://in.reuters.com/article/usa-navy-missiles-idINL2N16Q18P. (accessed May 9, 2019).
23. "J-15 chief Architect: 3-D Printing used in Developing New Fighter Jet" (March 4, 2013), *3ders.org*, retrieved from: http://www.3ders.org/articles/20130304-j-15-chief-architect-3d-printing-used-in-developing-new-fighter-jet.html. (accessed May 9, 2019).
24. Heyns, (May 23, 2013). "Lethal Autonomous Robotics", *UNIDIR Conference,* retrieved from: www.unidir.org. (accessed May 9, 2019).
25. Liran, A. (September 13, 2013). "Who Will Stop The Robots?", *Military and Strategic Affairs*, 5:2. See also: "Weaponization of Increasingly Autonomous Technologies", UNIDIR 6.
26. Heyns, *op. cit.*
27. Reddy, S.R. (June 22, 2016). "India and the Challenge of Autonomous Weapons," Paper by Carnegie India". *Carnegie*, retrieved from: https://carnegieindia.org. (accessed May 9, 2019).
28. Siboni, (January 4, 2014). "Dilemmas in Use of Autonomous Weapons", *INSS, TelAviv University Gabi.*
29. Lele, A. (October 30, 2015). "Wearable Military Technologies", *International Strategic and Security Studies Programme Reflection No. 30*, retrieved from: http://isssp.in/wearable-military-technologies/.
30. "An introduction to the Conference", *United Nations Office at Geneva.*
31. "Background on Lethal Autonomous Weapons Systems in the CCW", *United Nations Office at Geneva*, retrieved from: https://www.unog.ch/80256EE600585943/tpPages)/8FA3C2562A60FF81C1257CE600393DF6?OpenDocument. See also, "Autonomous weapons are a game-changer" (January 25, 2018), *The Economist*, retrieved from: https://www.economist.com/special-report/2018/01/25/autonomous-weapons-are-a-game-changer. (accessed May 9, 2019).

32. Canning, J., Riggs, G., Holland, O. (August 2004). "A Concept for the Operation of Armed Autonomous Systems on the Battlefield", Proc. AUVSI 2004.
33. Asaro, P. (December 2006). "What Should We Want From a Robot Ethic?", *International Review of Information Ethics*, Vol. 6.
34. Walzer, M. (2004). "Arguing About War", *Yale University Press.*
35. Goodwin, B. (August 20, 2012). "An HPC/UQ Front End is Key to Additive Manufacturing" (briefing, Additive Manufacturing Symposium, Washington, DC), *Lawrence Livermore National Laboratory.*
36. Houser, N. (2018). "Defense and National Security Industry Outlook: The future of defense and national security", *Deloitte*, retrieved from: https://www2.deloitte.com/global/en/pages/public-sector/articles/gx-national-security-industry-outlook.html. (accessed May 9, 2019).

Chapter 13

1. Clifford, C. (December 18, 2017). "*9 of the Most Jaw-dropping Things Elon Musk Said About Robots and AI in 2017"*. CNBC, retrieved from: https://www.cnbc.com/2017/12/18/9-mind-blowing-things-elon-musk-said-about-robots-and-ai-in-2017.html (accessed June 18, 2019).
2. Ibid.
3. Ibid.
4. "AI Policy – United States", *Future of Life Institute*, retrieved from: https://futureoflife.org/ai-policy-united-states/. (accessed June 17, 2019.
5. Delaney, J.K. (Congressman) (May 20, 2018). "France, China, and the EU All Have an AI Strategy. Shouldn't the US?" *Wired*, retrieved from: https://www.wired.com/story/the-us-needs-an-ai-strategy/ (accessed June 19, 2019).
6. Future of Life Institute, op.cit.
7. "AI Policy – Russia", *Future of Life Institute*, retrieved from: https://futureoflife.org/ai-policy-united-states/ (accessed June 17, 2019).
8. Ibid
9. Ibid.
10. Ibid.
11. "AI Policy – China", *Future of Life Institute*, retrieved from: https://futureoflife.org/ai-policy-united-states/ (accessed June 19, 2019).
12. Vincent, J. (March 14, 2019). "China About to Overtake America in AI Research", *The Verge*, retrieved from: https://www.theverge.com/2019/3/14/18265230/china-is-about-to-overtake-america-in-ai-research (accessed June 17, 2019).
13. Ibid.

14. Ibid.
15. Ibid.
16. "George W. Bush signs the Patriot Act – HISTORY" (Nov 16, 2009), *History.com* https://www.history.com/this-day-in-history,george-w-bush-signs-the-patriot-act (accessed February 07, 2019).
17. Gibbs, S. (July 17, 2017). "Elon Musk: Regulate AI to Combat 'Existential Threat' Before It's Too Late", *The Guardian*, retrieved from: https://www.theguardian.com/technology/2017/jul/17/elon-musk-regulation-ai-combat-existential-threat-tesla-spacex-ceo (Accessed 7 Feb. 2019).
18. Chaslot, G. (February 01, 2018). "How Algorithms Can Learn to Discredit the Media", *Medium*, retrieved from: https://medium.com/.@guillaumechaslot/how-algorithms-can-learn-to-discredit-the-media-d1360157c4fa (accessed February 07, 2019).
19. Gerstner, M.E. (1993). "Comment, Liability Issues with AI and Expert Systems", *Santa Clara Law Review,* Vol. 33, p. 239, retrieved from: http://digitalcommons.law.scu.edu/lawreview/vol33/iss1/7
20. United States v. Athlone Indus, Inc., 746 F. 2d 977,979 (3rd Circuit, 1984).
21. Masse, E. & Hidvegi, F. "Artificial Intelligence: What are the Issues for Digital Rights?", *Access Now*, retrieved from: https://www.accessnow.org/artificial-intelligence-issues-digital-rights/ (accessed December 01, 2018).
22. "Data Localization in a Globalised World: An Indian Perspective" (November 18, 2018), *The Dialogue.*
23. "Edward Snowden: Leaks that Exposed US Spy Programme" (January 17, 2014), *BBC News,* retrieved from: https://www.bbc.com/news/world-us-canada-23123964 (accessed October 30, 2018).
24. Bhat, A. & Katarki, S. (September 17, 2018) "India: The Debate – Data Localization And its Efficacy", *Mondaq,* retrieved from: http://www.mondaq.com/india/x/736934/Data+Protection+Privacy/The+Debate+Data+Localization+And+Its+Efficacy (accessed Oct 26, 2018).
25. "Data Protectionism: the Growing Menace to Global Business" (May 20, 2018), *Today Online*, retrieved from: https://www.todayonline.com/world/data-protectionism-growing-menace-global-business (accessed 26 Oct 2018).
26. Bauer, M., Ferracane, M.F., Lee-Makiyama, H., Van der Marel, E. (December 2016). "Unleashing Internal Data Flows in the EU: An Economic Assessment of Data Localisation Measures in the EU Member States", *ECIPE,* retrieved from: https://ecipe.org/publications/unleashing-internal-data-flows-in-the-eu/ (accessed June 19, 2019).

27. Bauer, M., Ferracane, M.F., Van der Marel, E. (30 May 2016). "Tracing the Economic Impact of Regulations on the Free Flow of Data and Data Localization.", *CIGI*, retrieved from: https://www.cigionline.org/sites/default/files/gcig_no30web_2.pdf. Accessed 26 Oct. 2018 (accessed June 19, 2019).

28. Andresen, Z. (2019). "Barriers to Digital Import-Export Trade Data Localization", *American Express*, retrieved from: https://www.americanexpress.com/us/foreign-exchange/articles/data-localization-affects-digital-import-export-trade/ (accessed November 15, 2018).

29. Bauer, Ferracane and van der Marel, *op. cit.*

30. "GSMA: Free Flow of Data Across Borders Essential for Asia's Digital Economies" (September 5, 2018), *GSMA*, retrieved from: https://www.gsma.com/newsroom/press-release/gsma-free-flow-of-data-across-borders-essential-for-asias-digital-economies/ (accessed November 15, 2018).

31. "Framework on personal data protection", *ASEAN*, retrieved from: http://asean.org/storage/2012/05/10-ASEAN-Framework-on-PDP.pdf (accessed November 15, 2018).

32. Myers, A. (January 19, 2016). "Top 10 operational impacts of the GDPR: Part 4 - Cross-border data transfer", *International Association of Privacy Professionals*, retrieved from: https://iapp.org/news/a/top-10-operational-impacts-of-the-gdpr-part-4-cross-border-data-transfers/ (accessed November 15, 2018).

33. Carrera, S., et.al. (2015). "Access to Electronic Data by Third-Country Law Enforcement Authorities: Challenges to EU Rule of Law and Fundamental Rights", *CEPS,* retrieved from: https://www.ceps.eu/ceps-publications/access-electronic-data-third-country-law-enforcement-authorities-challenges-eu-rule-law/ (accessed November 02, 2018).

34. UN Secretary General, (April 09, 2017), "Group of Governmental Experts on Developments in the Field of Information and Telecommunications in the Context of International Security", *UN Digital Library*, retrieved from: http://digitallibrary.un.org/record/799853?ln=en. (accessed November 15, 2018).

35. Cordis Europa, *"Managing Alternatives for Privacy, Property and Internet Governance" European Commission*, retrieved from: https://cordis.europa.eu/project/rcn/111214_en.html. (accessed November 15, 2018).

36. "Draft Legal Instrument on Government Led Surveillance and Privacy" (January 10, 2018), *OHCHR*, retrieved from: http://www.ohchr.org/Documents/Issues/Privacy/DraftLegalInstrumentGovernmentLed.pdf. (accessed November 15, 2018).

37. King, B., *op. cit.*, p. 64.
38. Clayton, C., (1997), "The Innovators Dilemma", *Harvard Business Review*; See also: Ibid p. 65
39. Sheng (2018), *op. cit.*
40. King, B., *op. cit.*, p. 67.
41. PricewaterhouseCoopers (June, 2016) "Central Banking 2020: Ahead of the Curve", *PwC*, retrieved from: https://www.pwc.com/gx/en/financial-services/pdf/cb-2020-ahead-of-the-curve.pdf. (accessed June 14, 2019).
42. King, B., *op cit.,* pp 79-80.
43. He, D. (June 2018) "Monetary Policy in the Digital Age", *International Monetary Fund*, retrieved from: https://www.imf.org/external/pubs/ft/fandd/2018/06/central-bank-monetary-policy-and-cryptocurrencies/he.htm. (accessed June 14, 2019).
44. PwC Report (2016), *op. cit.*
45. Lagarde, C. (September 29, 2017). "Central Banking and FinTech: A Brave New World?" *Speech at Bank of England Conference, London*, retrieved from: https://www.imf.org/en/News/Articles/2017/09/28/sp092917-central-banking-and-fintech-a-brave-new-world. (accessed June 14, 2019).
46. Chapter V, (June 2018). "Annual Economic Report 2018", *Bank of International Settlements*, retrieved from: https://www.bis.org/publ/arpdf/ar2018e.pdf. (accessed June 14, 2019).
47. Wilson, C. (2019). "Cryptocurrencies: The Future of Finance?" *Contemporary Issues in International Political Economy*, pp 359-394.
48. Bank of International Settlements, *op. cit.*
49. Carstens, A. (March 12, 2019), "The New Role of Central Banks", *Banks of International Settlements*, retrieved from: https://www.bis.org/speeches/sp190314.htm. (accessed June 14, 2019).
50. Lagarde, C. (2019), *op. cit.*
51. Lagarde, C. Ibid.
52. All quotes above are from Lagarde, C. (2019)
53. Lagarde, C (June 8, 2019). "The next steps in International Cooperation in Fintech" Opening Remarks by Christine Lagarde, Managing Director, IMF G20 High Level Seminar "Our Future in the Digital Age" Fukuoka, Japan, retrieved from: https://www.imf.org/en/News/Articles/2019/06/07/sp060819-lagarde-the-next-steps-for-international-cooperation-in-fintech
54. Ibid.

Chapter 14

1. Speech of Nirmala Sitharaman, Minister of Finance, Government of India, while presenting the Budget 2019-2020 in the Indian Parliament (July 5, 2019).
2. For a lucid explosion of how the rapid technological advances are affecting day-to-day lives in India, *See* Karnik, K. (2018). *Evolution: Decoding India's Disruptive Tech Story,* New Delhi: Rupa Publications.
3. Peri D. (August 24, 2018). "ISRO telemedicines nodes for soldiers in high-altitude area", *The Hindu*, retrieved from: https://www.thehindu.com/news/national/isro-telemedicine-nodes-for-soldiers-in-high-altitude-areas/article24774110.ece. (accessed May 20, 2019).
4. Karnik, *op. cit.*, p.2
5. "UPI Product Statistics", *National Payments Corporation of India*, retrieved from: https://www.npci.org.in/product-statistics/upi-product-statistics (accessed May 20, 2019.
6. "Digital Payment Industry in India" (January 04, 2019), *IBEF Blog, Perspectives on India*, retrieved from: https://www.ibef.org/blogs/digital-payment-industry-in-india. (accessed May 20, 2019
7. Mukherjee, S. (February 15, 2018). "Digital Payments to evolve into a $1 Trillion Behemoth in India by 2023: Credit Suisse", *Inc 42*, retrieved from: https://inc42.com/buzz/digital-payments-credit-suisse-report/ (accessed May 20, 2019).
8. "National Strategy for Artificial Intelligence – Discussion Paper" (June 2018), *NITI Aayog*, p.33, retrieved from: https://niti.gov.in/writereaddata/files/document_publication/NationalStrategy-for-AI-Discussion-Paper.pdf, (accessed May 20, 2019).
9. "The Right to Privacy in the Digital Age" (June 30, 2014), *Report of the Office of the United Nations High Commissioner for Human Rights, UNGA.* See also, Art. 12, UN General Assembly, Universal Declaration of Human Rights, 10 December 1948, 217 A (III); Art. 17 UN General Assembly, International Covenant on Civil and Political Rights, 16 December 1966, United Nations, Treaty Series, vol. 999, p. 171,
10. Slay, J. & Miller, M., "Lessons Learned from the Maroochy Water Breach: Critical Infrastructure Protection", *International Federation for Information Processing*, 253, pp. 73-82, Springer Publications.

INDEX

THE DIALOGUE

The Dialogue is a research driven think tank working on the intersection of technology, society and public policy. Founded in June 2017, The Dialogue drives informed discourses towards a progressive policy ecosystem in India.

Our aim is to enable a more coherent policy discourse in India backed by evidence and layered with the passion to transform India's growth, to help inform on public policies, analyze the impact of governance and subsequently, develop robust solutions to tackle our challenges and capitalize on our opportunities. Towards this, we deliver research projects and informed discourses around issues concerning technology, trade foreign policy and development sectors.

Founded by a group of young citizens, we are committed to contribute in the development of a New India, based on the pillar of national interest. Moreover, in recent times, The Dialogue has been at the forefront of the policy discourse on emerging technologies. We have been a key voice in the larger policy discussions and have closely engaged with the government and other stakeholders on matters of such policy areas.

The Dialogue was established on the principles of taking India forward, which has been the driving force of our work, and we intend to play an important role towards supporting India's growth.

The Dialogue
Inform Engage Ideate